Samuel Edward Warren

# The Elements of descriptive Geometry, Shadows and Perspective

Samuel Edward Warren

**The Elements of descriptive Geometry, Shadows and Perspective**

ISBN/EAN: 9783337188030

Printed in Europe, USA, Canada, Australia, Japan

Cover: Foto ©berggeist007 / pixelio.de

More available books at **www.hansebooks.com**

# THE ELEMENTS

OF

# DESCRIPTIVE GEOMETRY,

SHADOWS AND PERSPECTIVE.

WITH A BRIEF TREATMENT OF

TRIHEDRALS, TRANSVERSALS, AND SPHERICAL, AXONOMETRIC, AND OBLIQUE PROJECTIONS.

*FOR COLLEGES AND SCIENTIFIC SCHOOLS.*

BY

S. EDWARD WARREN, C.E.,

FORMERLY PROFESSOR IN THE RENSSELAER POLYTECHNIC INSTITUTE,
ETC., ETC.

NEW YORK:
JOHN WILEY & SONS, PUBLISHERS,
15 ASTOR PLACE.
1877.

Copyright, 1877,
JOHN WILEY AND SONS.

S. W. GREEN,
PRINTER AND ELECTROTYPER,
16 and 18 Jacob Street.

# CONTENTS.

|  | PAGE |
|---|---|
| PREFACE. | |
| DESCRIPTIVE GEOMETRY.—Definitions. General principles. Fundamental problems. | 3 |
| *Preliminary* | 3 |
| *Definitions and first principles* | 4 |
| *Fundamental problems* | 7 |

### A—Projections.

PROBLEM I.—To make the projections of points in various positions .. 7
PROBLEM II.—To draw the projections of lines in various positions in space .................................................................. 10

### B—Tangencies.

PROBLEM III.—To draw lines having various positions in the planes of projection .................................................................. 12

### C—Intersections.

PROBLEM IV.—To represent planes in various positions.............. 13

### D—Development.

PROBLEM V.—To revolve a point so as to immediately show the form and extent of its path................................................ 16
*Classification of surfaces and lines*............................... 17

---

# DESCRIPTIVE GEOMETRY.

## BOOK I.—SURFACES OF REVOLUTION.

### PART I.—RULED SURFACES.

#### CHAPTER I.

##### GENERAL PROBLEMS OF THE POINT, LINE, AND PLANE.

###### A—Projections.

PROBLEM VI.—Having given two projections of a point and a line, to find their projections on any new planes of projection............ 20

## B—Tangencies.

PROBLEM VII.—To draw a straight line in a plane which is given by its traces; and a plane to contain a line which is given by its traces.. 21

## C—Intersections.

PROBLEM VIII.—To find the traces of a line given by its projections... 22
THEOREM I.—The projection of a right angle will be a right angle when at least one of its sides is parallel to the plane of projection 24
THEOREM II.—If a line is perpendicular to a plane, its projections will be perpendicular to the traces of the plane.................. 25
PROBLEM IX.—To construct the traces of a plane under frequently recurring conditions............................................. 25
    1°.—Through three given points............................ 25
    2°.—Through a given point and line........................ 26
    3°.—Through two given parallel or intersecting lines......... 26
    4°.—Through one given line and parallel to another.......... 26
    5°.—Through a given point and perpendicular to a line....... 26
    6°.—Through a given point and parallel to any two given lines. 26
    7°.—Through a given point and parallel to a given plane..... 26
PROBLEM X.—To construct the intersection of two planes, given by their traces........................................................ 28
*The general case*................................................. 28
*Particular cases.*—1°. The traces of the intersection inaccessible........ 29
              2°. The planes nearly perpendicular to the ground line......................................................... 29
PROBLEM XI.—To construct the intersection of a line with a plane.... 30
*The general case*.................................................. 30
*Particular cases.*—1°. The given line perpendicular either to H or V.... 31
              2°. The plane given by other lines than its traces.... 32

## D—Development.

*General methods of solution.*—By fixed magnitudes. By rotations. By change of planes................................................. 33
PROBLEM XII.—To find the true length of a line joining two given points............................................................ 34
   *First method.*
   *Second method.*
PROBLEM XIII.—To find the shortest distance from a given point to a given line. ................................................... 35
PROBLEM XIV.—To find the shortest distance between two lines not in the same plane................................................. 36
PROBLEM XV.—To find the true size of the angle between two given lines............................................................. 38
PROBLEM XVI.—To bisect a given angle in space. 1°. General case.. 39
                              2°. Special case.. 39

## CONTENTS.

PAGE

PROBLEM XVII.—To find the true size of the angle included between two given planes .................... 39
    1°.—One of the given planes a plane of projection ............ 40
    2°.—The general case without auxiliary developments ........ 41
    3°.—The general case with auxiliary developments ........... 41
PROBLEM XVIII.—To find the angle made by a line with a plane ...... 42
PROBLEM XIX.—To reduce an angle to the horizon ................ 43
PROBLEM XX.—To construct the projections of a plane figure lying in a given plane ................................................ 43
    THEOREM III.—The projections of a circle seen obliquely are ellipses ................................................ 45

## CHAPTER II.
### DEVELOPABLE SURFACES.

*Definitions* .................................................. 46

#### A.—Projections.

PROBLEM XXI.—To construct an ellipse under various given conditions. 47
    1°.—By radials from the extremities of an axis ............... 47
    2°.—By concentric circles on the axes ....................... 48
    3°.—On a given pair of conjugate diameters .................. 48
        *First.* By the oblique projection of a circle ................ 48
        *Second.* By the oblique projection of the radial method ...... 49
PROBLEM XXII.—To construct tangents to a given ellipse ............ 50
    1°.—At a given point on the curve ......................... 50
    2°.—Through a given exterior point ......................... 50
    3°.—Parallel to a given line ............................... 51
PROBLEM XXIII.—To represent a cone of given axis and diameter by the projections of its vertex, circular section, and visible limits; the axis being oblique to both H and V; also any element or point of the surface ................................................ 51
    1°.—Four points of the section ............................. 52
    2°.—The shorter axis of the section ......................... 52
    3°.—Other points of the section ............................ 53
    4°.—Partial plane construction of the section ................ 53
    5°.—Visibility ........................................... 53
*The conic sections in general* ................................... 54
    THEOREM IV.—In the ellipse, section of a cone made by a plane actually cutting all its elements, the sum of the distances from any point of the curve to two fixed points within it is constant and equal to the transverse axis ............................ 56
PROBLEM XXIV.—To represent a cone by the projections of its vertex, horizontal trace, and extreme elements ..................... 58
    1°.—Without the conjugate axis ............................ 58
    2°.—By the axes of the trace .............................. 58

## CONTENTS.

PAGE

PROBLEM XXV.—To construct the projections of a cylinder of revolution having a given horizontal trace or base; also any element and point of the surface .................................................. 59

PROBLEM XXVI.—To construct the projections of a cone of revolution having a given elliptical trace or base...................... 60

PROBLEM XXVII.—To construct the projections of a cone whose axis is parallel to the ground line, and of any of its elements and points ............................................................. 61

### B—Tangencies.

PROBLEM XXVIII.—To construct a plane, tangent to a cone on a given element .......................................................... 64

PROBLEM XXIX.—To construct a plane parallel to a given line, and tangent to a cylinder whose axis is parallel to the ground line....... 65

PROBLEM XXX.—Through a given point in space, to construct a tangent plane to a cone whose horizontal trace is given ............... 66

PROBLEM XXXI.—To pass a plane parallel to a given line, and tangent to a cylinder whose axis is oblique to H and V, and one of whose traces is given ............................................... 67

PROBLEM XXXII.—To construct a plane parallel to a given line and tangent to a cone whose horizontal trace is known, and whose axis is oblique to both H and V......................................... 68

### C—Intersections.

PROBLEM XXXIII.—To find the intersection of a plane perpendicular to V with a cylinder whose axis is oblique to both planes of projection, and whose horizontal trace is given ........................... 72

PROBLEM XXXIV.—To find the intersection of a plane with a cylinder, both being oblique to both planes of projection............... 73

PROBLEM XXXV.—To find the intersection of an oblique plane with a cone whose axis is vertical ........................................ 74
    1°.—The points on the extreme elements..................... 75
    2°.—The highest and lowest points .......................... 75
    3°.—Points on the foremost and hindmost elements.......... 75
    4°.—Points on any other elements........................... 76
    5°.—Points found by horizontal auxiliary planes ............ 76

PROBLEM XXXVI.—To find the intersection of a plane and a cone, both of which are oblique to both planes of projection............... 76

PROBLEM XXXVII.—To find the intersection of two cones, the axis of each being perpendicular both to H and V...................... 78
    1°.—Selection of planes and construction of points............ 78
    2°.—Connection of points .................................. 79
    3°.—Number and disposition of curves....................... 80
    4°.—Visibility of the curve................................. 80
    5°.—Visibility of the extreme elements ..................... 80

PROBLEM XXXVIII.—To find the intersection of two cylinders whose axes intersect at right angles, one of them being parallel to the ground line...... 81
PROBLEM XXXIX.—To construct the tangent line at a given point of any plane or double-curved intersection............ 82

### D—Development.

PROBLEM XL.—To find the true form and size of the intersection of any cylinder or cone by a plane, with the tangent to the revolved curve............ 84
    1°.—A plane and cylinder, the plane perpendicular to V....... 84
    2°.—A plane and cylinder, both oblique to H and V.......... 84
    3°.—A plane and cone, both oblique to H and V............. 84
PROBLEM XLI.—To develop the convex surface of a cylinder, together with its intersection with a plane, or with another cylinder or a cone, and any tangent line to the curve........... 85
PROBLEM XLII.—To develop the convex surface of a cone of revolution, with any curve on its surface, and a tangent to the curve..... 86

## CHAPTER III.

### WARPED SURFACES.

THEOREM V.—The surface generated by the revolution of a straight line around another straight line not in the same plane with it is warped............ 89
THEOREM VI.—The warped surface of revolution has two sets of elements, such that at every point of the surface two elements can be drawn, one of each set........... 90

### A—Projections.

PROBLEM XLIII.—To construct the projections of the warped surface of revolution............ 91
PROBLEM XLIV.—To represent the vertical projection of the warped surface of revolution by its meridian, parallel to V................. 92
    1°.—By symmetrical elements............ 93
    2°.—By horizontal circles ............ 93
THEOREM VII.—The meridian curve of the warped surface of revolution is a hyperbola............ 93
PROBLEM XLV.—Having given either projection of a point on a warped hyperboloid of revolution, to find its other projection........ 95

### B—Tangencies.

THEOREM VIII.—Every tangent plane to a warped hyperboloid of revolution is also a secant plane containing two elements, and is tangent at their point of intersection........................ 96
PROBLEM XLVI.—To construct a tangent plane at a given point of the warped hyperboloid of revolution ........................... 97
PROBLEM XLVII.—To construct a plane, tangent to a warped hyperboloid of revolution, and containing a given line.................. 98

### C—Intersections.

PROBLEM XLVIII.—To find the intersection of a plane and warped hyperboloid of revolution, and a tangent to the curve at a given point. 99
    1°.—By auxiliary planes........................................ 99
    2°.—By auxiliary cones......................................... 99
    3°.—The tangent line........................................... 101

### D—Development.

## PART II.—DOUBLE-CURVED SURFACES.

### A—Projections.

PROBLEM XLIX.—To represent the projections of any double-curved surface of revolution........................................ 105
PROBLEM L.—Having one projection of a point upon a double-curved surface, to find its other projection ......................... 107

### B—Tangencies.

PROBLEM LI.—To construct a plane, tangent to an annular torus at a given point of contact .................................................. 109
PROBLEM LII.—To construct a tangent plane to a sphere, and through a given line, by the direct method......................... 111
PROBLEM LIII.—To construct a plane, tangent to a sphere and through a given line, by means of an auxiliary tangent cone ........ 112
PROBLEM LIV.—To construct a plane through a given line and tangent to a given double-curved surface of revolution, by means of a concentric auxiliary surface................................................. 114
PROBLEM LV.—To construct a plane through a point in space and tangent to a double-curved surface of revolution on a given section of the surface ................................................................. 115
    1°.—A parallel................................................. 115
    2°.—A meridian................................................ 116

## C—Intersections.

PROBLEM LVI.—To find the intersection of a double-curved surface of revolution with a plane............................................. 117
    I.—That of an oblate spheroid ............................. 117
        1°.—The method of meridian........................... 117
        2°.—The method of parallels........................... 118
    II.—The plane sections of the annular torus.................. 118
THEOREM IX.—The section of an annular torus, made by a bi-tangent plane to it, consists of a pair of circles whose diameter is that of the generating circle plus that of the gorge .................. 119
PROBLEM LVII.—To find the intersection of a single-curved surface with a double-curved surface, both of revolution .................... 120
PROBLEM LVIII.—To find the intersection of two double-curved surfaces whose axes intersect ............................................. 122
PROBLEM LIX.—To find points of the intersection of ellipsoids of revolution whose axes are not in the same plane.................... 123
PROBLEM LX.—To construct a tangent line at a given point of the intersection of two surfaces of revolution, one or both of which is double-curved ...................................................... 123
        1°.—To the intersection of the plane and annular torus......... 124
        2°.—To the intersection of the ellipsoid and paraboloid........ 124

# BOOK II.—SURFACES OF TRANSPOSITION.

## PART I.—RULED SURFACES.

### CHAPTER I.

#### PLANE SURFACES.

##### A—Projections.

PROBLEM LXI.—To draw lines through a given point, which shall determine a plane parallel to a plane which is given by means of two lines in it................................................................ 126

##### B—Tangencies.

PROBLEM LXII.—To find the traces of a plane which is given by two lines, neither of which intersects the planes of projection within convenient limits.................................................... 127

## C—Intersections.

PROBLEM LXIII.—To find the intersection of two planes, each of which is given by its horizontal trace and one point............ 128
    THEOREM I.—Of transversals................................... 129
PROBLEM LXIV.—To find the intersection of two planes whose traces do not meet within the limits of the figure.............. 129
    THEOREM II.—Of transversals.................................. 130
    THEOREM III.—Of transversals................................. 131
    THEOREM IV.—Of transversals.................................. 131
PROBLEM LXV.—Having given two lines and a point, to construct a third line through this point so that the three lines shall all meet in one point................................................... 131
    THEOREM V.—Of transversals................................... 132
PROBLEM LXVI.—Through a given point to draw a line that shall intersect two given lines.............................................. 132
    *First solution.*
    *Second solution.*
    THEOREM VI.—Of transversals.................................. 133

# CHAPTER II.

### DEVELOPABLE SURFACES.

*Definitions, etc.*.................................................. 134
PROBLEM LXVII.—To construct the projections of a helix and of its tangent............................................................ 137
PROBLEM LXVIII.—To construct the projections of a developable helicoid, its axis being vertical................................... 138
*Examination of the horizontal trace of the helicoid.*............... 140

### B—Tangencies.

PROBLEM LXIX.—To construct a tangent plane along a given element of a developable helicoid........................................ 141
PROBLEM LXX.—To construct a tangent plane to a developable helicoid, through a given exterior point............................. 141
PROBLEM LXXI.—To construct a tangent plane to a developable helicoid and parallel to a given line.................................. 141

### C—Intersections.

*Developable helicoids.*............................................. 143
*General view of the intersections of cones. Infinite branches.*..... 144
    I.—Varieties depending on the relation of the tangent auxiliary planes........................................................ 144
    II.—Varieties depending on the relation of either cone, $V$, to one composed of elements through its vertex, and parallel to those of the other cone $V_1$........................ 144

PROBLEM LXXII.—To find the intersection of two cones when the curve has infinite branches.................................................. 146

### D—Development.

PROBLEM LXXIII.—To develop the convex surface of a cylinder of transposition whose elements are oblique to both planes of projection 148
PROBLEM LXXIV.—To develop a helix by means of its curvature..... 149
PROBLEM LXXV.—To develop so much of the surface of a developable helicoid as lies between two planes perpendicular to its axis.... 150

## CHAPTER III.

### WARPED SURFACES.

*Principles*.......................................................... 153
THEOREM X.—The hyperbolic paraboloid is doubly ruled, or has two sets of elements................................................ 155

### A—Projections.

PROBLEM LXXVI.—To construct the projections of a hyperbolic paraboloid having given the directrices and plane director of one generation .................................................................. 157
PROBLEM LXXVII.—Having one projection of a point on the surface of a hyperbolic paraboloid, to find its other projection.............. 158
PROBLEM LXXVIII.—Having given a hyperbolic paraboloid, by a pair of elements of each generation, and having also a point of the surface, to construct an element of either generation through that point...... 159
PROBLEM LXXIX.—To construct the projections of a warped elliptical hyperboloid..................................................... 160
PROBLEM LXXX.—To construct the projections of an oblique helicoid ................................................................. 161
PROBLEM LXXXI.—To construct the projections of a conoid......... 163
PROBLEM LXXXII.—To construct the projections of a warped arch... 163
PROBLEM LXXXIII.—To construct elements of a general warped surface having three given directrices ................................. 165
PROBLEM LXXXIV.—To construct elements of a general warped surface having two directrices and a plane director .................... 166

### B—Tangencies.

PROBLEM LXXXV.—To construct the tangent plane at a given point of a hyperbolic paraboloid........................................ 170
PROBLEM LXXXVI.—To construct the tangent plane at a given point of an elliptical hyperboloid...................................... 170
PROBLEM LXXXVII.—To construct the tangent plane at a given point of a warped surface by the direct method—that is, without an auxiliary raccording surface ................................................ 171

PROBLEM LXXXVIII.—To construct a tangent plane to a warped surface indirectly—that is, by means of an auxiliary raccording surface.................................................................. 172
    1°.—To a conoid.................................................. 172
    2°.—To the warped arch........................................ 172

### C—Intersections.

THEOREM XI.—The intersection of a right conoid having an elliptical directrix, with any plane parallel to that directrix, is an ellipse.. 175
THEOREM XII.—The intersection of an oblique helicoid by a plane perpendicular to its axis, is a spiral of Archimedes................ 175

## PART II.—DOUBLE-CURVED SURFACES OF TRANSPOSITION.

### A—Projections.

PROBLEM LXXXIX.—To construct the projections of a serpentine.... 178

### B—Tangencies.

PROBLEM XC.—To construct a plane through a given line and tangent to an ellipsoid of three unequal axes ........................ 179

### C—Intersections.

PROBLEM XCI.—To construct the projections of a spherical epicycloid, and of a tangent line to it at any point ............................ 183
PROBLEM XCII.—To find the intersection of a cone and sphere, the centre of the sphere being at the vertex of the cone, together with a tangent to the curve................................................ 184

### D—Development.

PROBLEM XCIII.—To develop the surface of an oblique cone......... 186

# SHADES AND SHADOWS.

*First Principles*..................................................... 188

### A—Shades and Shadows on Planes.

PROBLEM I.—To find the shades and shadows on a rectangular block having a panel in front and a tablet on top; all its edges being parallel or perpendicular to H or V............................. 190
PROBLEM II.—To find the shades and shadows of a stone cross........ 193
PROBLEM III.—To find the shadow of a turnstile on H and on a vertical plane oblique to V........................................ 194

|                                                                                               PAGE |
|---|

+PROBLEM IV.—To find the shadow cast on a flight of steps by one of its piers .................................................... 195
       1°.—Preliminary ................................................ 195
       2°.—The shadows of $AB - A'B'$ ................................. 196
       3°.—The remaining shadows on step 1 ........................ 196
       4°.—The remaining shadows ................................... 196
       5°.—Construction from the side elevation .................... 196
PROBLEM V.—To find the shadows of chimneys on side and end roof surfaces ............................................................ 198
       1°.—Of the left-hand chimney ................................. 199
       2°.—Of the right-hand chimney ............................... 199

### B—Shades and Shadows on Developable Surfaces.

PROBLEM VI.—To find the shadow of a rectangular abacus upon a cylindrical column, and the shade of the column .................. 201
PROBLEM VII.—To find the shadow of the upper base of a vertical hollow half cylinder upon the visible interior surface ............. 203
PROBLEM VIII.—To find the shadow of a vertical cylindrical turret upon a concave cylindrical roof surface ......................... 203
       1°.—The shadow of an element of shade of the turret ........ 204
       2°.—The shadow of the upper base of the turret ............. 204
PROBLEM IX.—To find the shadow of a square abacus upon a conical column, and the shade of the column ............................ 205
       1°.—The shade of the column .................................. 205
       2°.—The point in the meridian plane of rays ................. 206
       3°.—The shadow of the left-hand edge ....................... 206
       4°.—The highest point of the shadow ........................ 206
       5°.—The point of shadow on the element of shade ........... 206
       6°.—Indirect construction of points ........................... 206
PROBLEM X.—To find the shadow of a vertical cylinder upon an upright cone ................................................................ 207

### C—Shades and Shadows on Warped Surfaces.

PROBLEM XI.—To find the shades and shadows on a triangular threaded screw whose axis is vertical ............................... 210
       1°.—A point of shade on the inner helix, by declivity cones ... 210
       2°.—A point of shade on the outer helix, by helical transposition 211
       3°.—The shadow of the curve of shade on the upper surface of a thread ..................................................... 212
       4°.—The shadow of the outer helix on the thread below ....... 213

### D—Shades and Shadows on Double-curved Surfaces.

PROBLEM XII.—To find the curve of shade upon a sphere ............ 214
PROBLEM XIII.—To find the curve of shade on a torus ............... 215
       *Direct methods.*

|  | PAGE |
|---|---|
| 1°.—On the apparent contours | 215 |
| 2°.—The highest and lowest points | 216 |
| 3°.—Intermediate points by tangent planes of rays | 216 |
| *Indirect methods.* 1°.—By circumscribed cones | 217 |
| 2°.—By circumscribed spheres | 217 |
| PROBLEM XIV.—To find the shadow of a niche on itself | 218 |
| 1°.—By direct construction | 219 |
| 2°.—The indirect method | 220 |
| 3°.—The point on the curve of contact | 220 |
| PROBLEM XV.—To find the brilliant point of a double-curved surface | 221 |
| 1°.—On the sphere | 221 |
| 2°.—On a concave tower roof | 221 |

# LINEAR PERSPECTIVE.

*First Principles* .................................................... 223

### A—Perspective of Plane-Sided Bodies—Various Methods.

PROBLEM I.—To find the perspective of a monument composed of a square prism capped by a square pyramid ........................ 225
*Remarks on the method of three planes* ............................. 226
PROBLEM II.—To construct the perspective of a pier, with the points of convergence of the perspectives of its parallel lines ............. 227
PROBLEM III.—To find the perspective of a square block and of its shadows on the horizontal plane .................................. 229
PROBLEM IV.—To find the perspective of a shelf on a lateral wall, perpendicular to the perspective plane ............................ 230
*Indirect or artificial methods* ..................................... 231
PROBLEM V.—To find the perspective of a horizontal and of an oblique straight line: first, by visual rays; second, by trace and vanishing point; the vertical plane being also the perspective plane .... 234
PROBLEM VI.—To find the perspective of a rectangular prism, one face of which is in **H** and another in **V**; and to find a point of its shadow on **H** .................................................... 235
    THEOREM XIII.—The vanishing points of rays and of their horizontal projections are on the same perpendicular to the ground line .......................................................... 237
PROBLEM VII.—To find the perspective of a rectangular prism lying on **H**, but with its horizontal edges oblique to **V**; also its shadow on **H** ............................................................ 237
PROBLEM VIII.—To find the perspective of a wall whose front face is sloping, and whose horizontal edges are oblique to **V**; also its shadow on **H** .................................................... 239
PROBLEM IX.—To construct the perspective of an interior, containing various objects, by the method called that of scales ......... 241

PROBLEM X.—To find the perspectives of two circles lying in H, the centre of one of them being, with the eye, in a profile plane ......... 243

### B—Perspectives of Developable Surfaces.

PROBLEM XI.—To find the perspective of an elbow arch and of its shadows............................................................. 246
    1°.—The outlines ............................................... 246
    2°.—The groin.................................................. 247
    3°.—The shadows. Of the front circle on the perpendicular cylinder................................................... 248
    4°.—On the parallel cylinder.................................. 249

### C—Perspectives of Warped Surfaces.

### D—Perspectives of Double-curved Surfaces.

PROBLEM XII.—To find the perspective of the double-curved surface of revolution, called a piedouche or scotia ...................... 251
    1°.—The projections of the apparent contour ................. 251
    2°.—The curve of shade ...................................... 252
    3°.—The perspective ......................................... 252
    4°.—Method by perspectives of parallels ..................... 253
    *Adhemar's general method*........................................ 253
PROBLEM XIII.—To find the perspective of a point taken at pleasure in H...................................................................... 254
PROBLEM XIV.—To find the perspective of any quadrilateral in H..... 255
PROBLEM XV.—To find the perspective of a stone bridge............. 256
    1°.—The perspective plan ..................................... 257
    2°.—The perspective elevation ................................ 257
    3°.—The arch points........................................... 258

# SPHERICAL PROJECTIONS.

Orthographic ........................................................ 261
Stereographic........................................................ 262
Globular............................................................. 263
Equidistant, or approximate globular................................. 264
Cylindrical.......................................................... 264
Mercator's........................................................... 265
Gnomonic............................................................. 266
Polar ............................................................... 266
Conic ............................................................... 267
Polyconic ........................................................... 267
Equidistant polyconic ............................................... 269
Flamstead's ......................................................... 269

## TRIHEDRALS.

PROBLEM I.—Given the three sides of a spherical triangle—that is, the three plane angles of a trihedral—to find its angles—that is, the diedrals of the trihedral.................................................. 271

PROBLEM II.—Given two sides and the included angle of a spherical triangle—that is, two plane angles and the included diedral of a trihedral—to find the remaining parts..................................... 272

PROBLEM III.—Given two sides and an opposite angle of a spherical triangle—that is, two plane angles and the diedral opposite one of them, in a trihedral—to find the remaining parts.................... 272

---

## AXONOMETRIC AND OBLIQUE PROJECTIONS.

PROBLEM I.—Having given three axes, the axonometric projections of the three edges of a solid right angle, to find the scales of those edges—that is, the ratio of distances on them to their true size...... 276

PROBLEM II.—To construct the axonometric projection of a monument embracing prismatic and pyramidal portions...................... 277

PROBLEM III.—To construct the axonometric projection of a vertical cylinder capped by a hemisphere of less diameter. ............ 279

PROBLEM IV.—To construct comparative views of one given object, in elevation, axonometric and oblique projections .................. 280

## PREFACE.

THIS volume, entirely new, and not abridged from my larger ones, was prepared partly to meet an evident demand for brief text-books; however desirable it is that its subjects should be pursued in collegiate and professional scientific courses as fully as treated in those volumes.

Problems in shadows and perspective can, if preferred, be taken up in immediate connection with those of descriptive geometry, of which they are applications.

Perspective has been partly superseded by photography for representing existing structures, but is still indispensable for the pictorial representation of proposed structures, and for other purposes; besides being of peculiar value as a branch of geometrical study.

Spherical projections, Trihedrals, and "One plane descriptive," being of limited and special application, and found in various accessible works, I have treated the two former very briefly, after consultation with able advisers, and have omitted the latter.

The room thus gained has been devoted to some fresher topics, and useful extensions of the principal contents, including axonometric projections; a few examples of transversals from Olivier, as treated by descriptive geometry; Adhemar's general method of condensed perspective construction, and, with very slight enlargement, numerous points, making the work more complete in itself, and affording answers to such questions as interested students are always found ready to ask.

The tested and satisfactory feature of examples for practice under each problem has been retained; and, generally, I have sought to make my work as far as possible a clear addition to the resources of professors and students.

# DESCRIPTIVE GEOMETRY.

DEFINITIONS, GENERAL PRINCIPLES, FUNDAMENTAL PROBLEMS.

*Preliminary.*

1. GEOMETRY is variously divided:

1°. As to *grade*, into *lower* and *higher*. The former is limited to the conic sections and bodies bounded by them, and includes elementary geometry, or that of the straight line and circle and forms bounded by these. Higher geometry embraces all other magnitudes.

2°. As to *method*, into *analytic*, in which we proceed, by the aid of equations, from general to particular truths, and *synthetic*, in which we proceed, by the aid of diagrams, from particular to general truths.

3°. As to *dimensions*, into that of *two*, and of *three* dimensions, otherwise called *plane* and *solid* geometry.

4°. As to its *object*, into that of *ratio and measure*, and that of *form and position*.

2. *Illustration.*—The propositions, "The area of a triangle is the product of its base by half its altitude"; "Similar figures are to each other as the squares of their homologous sides"; "The volume of a sphere is two thirds of that of the circumscribed cylinder"; are those of *ratio and measure*, and are evidently of a different class from; "The lines from the vertices of any triangle to the middle points of the opposite sides all intersect at one point"; "The intersections of the opposite sides of any inscribed hexagon are in the same straight line"; "But one perpendicular can be drawn from

a given point to a given plane"; "A plane, perpendicular to two other planes, is perpendicular to their intersection"; which are propositions of *form and position*.

### Definitions and First Principles.

3. DESCRIPTIVE GEOMETRY is the geometry of form and position in space, treated by means of exact representations of magnitudes in space upon planes. These representations are called *projections*, and the planes on which they are made *planes of projection*.

By reason of the definite relation of projections to the objects represented by them, they constitute an exact *graphical description* of these objects, written, as is sometimes said, in the language of projections. Hence the propriety of the name, *descriptive* geometry.

4. *Descriptive geometry*, therefore, treats, *first*, of the exact representation of geometrical forms upon planes; thence, *second*, of the graphical solution, on planes, of problems upon these forms in space. To this, according to (3), is sometimes added, *third*, the demonstration, by the method of projections, of properties of form and position.

5. *Two planes of projection* (3) are commonly employed in descriptive geometry. They are usually perpendicular to each other, and may be viewed perpendicularly or obliquely, and from a finite or an infinite distance.

When the planes of projection are viewed from an infinite distance, the lines of vision are *parallel*, and are called *projecting lines*, and their intersections with the planes of projections are called the *projections* of the given points through which they are drawn.

When the planes are viewed from a finite distance, the lines of vision, then divergent, are called visual rays.

6. We thus have summarily:
I. *Parallel or cylindrical projection*, in which the point of

sight is at an *infinite* distance, and the projecting lines are *parallel.* This includes:

    1°. Perpendicular, orthogonal, or orthographic projection, when the planes of projection are viewed perpendicularly.

    2°. Oblique projection when they are viewed obliquely.

  II. *Radial, conical, or scenographic projection,* when the point of sight is at a finite distance. This is the same as Natural Perspective.

We shall first and principally consider the perpendicular projection, it being most generally useful.

7. Pl. I., Fig. 1, represents the planes of projection, briefly designated as **H** and **V**, one of them, *FHbh,* being horizontal, and thence called the *horizontal plane;* the other, *UVlv,* vertical, and thence called the *vertical plane.* Their intersection, indicated on the plates by the letters *GL,* expressed or understood, is called the *ground line,* from which each is supposed to extend each way indefinitely. They thus include *four diedral angles,* numbered as in the figure, the first containing the point of sight, and each is divided by the ground line into two parts: the horizontal plane into the *front* and *back* parts, *FHGL* and *bhGL;* and the vertical plane into the *upper* and *lower* parts, *UVGL* and *lvGL.*

8. *Coincidence of the planes* **H** *and* **V** *on paper.*—Since the actual solution of problems is effected on a single plane surface of paper, the latter is made to represent the two separate planes in space as follows: The vertical plane *UVlv,* Pl. I., Figs. 1–4–8, is revolved *backward,* or from the eye of the observer, and about the ground line *GL* as an axis, until its *upper* part, *UV,* coincides with the *back* part, *bh,* of the horizontal plane. Its lower part, *lv,* will consequently coincide with the front part, *FH,* of the horizontal plane. In this revolution, every point of the vertical plane not in the ground line describes a quadrant whose plane is perpendicular to the ground line; as at *Ub* or *lv,* Fig. 1.

It results from this revolution, that having drawn a line *GL*, Pl. I., Fig. 3, to represent the ground line, all that part of the paper which is in front of it represents both the front part of the horizontal, and the lower part of the vertical plane; while the part beyond *GL* represents both the upper part of the vertical, and the back part of the horizontal plane.

9. *Stages of Representation.*—These are four: *First.* Models, such as the student can prepare for home use by cutting two cards, each across half its width, so that they may be halved together, giving one vertical, as *UVlv*, Pl. I., Fig. 1, and the other horizontal, as *FHbh*. Needles, or threads, may then serve for lines, and bits of stiff paper for various planes.

*Second.* Pictorial figures, or pictures of models, like Pl. I., Figs. 1–4–8.* These are in some respects better than actual models, since the opacity of the latter hides parts which can very well be shown by dotted lines, or as if the model were transparent.

*Third.* Pictorial representations like Pl. I., Figs. 2–5–9, of the two planes, and their contents, after being made to coincide as described in (8).

In these it is evident that the objects in space cannot be represented as in Figs. 1–4–8.

*Fourth.* The actual constructions made as last described in (8). Pl. I., Fig. 3.

Of these four stages, the first three are only illustrative, and intended to assist the imagination to see in the projections, alone, the objects and planes of projection in their real positions in space. The second and third, especially the second, more freely used at first, or in explaining entirely new forms, are afterwards omitted.

Since the *ideal* solution of every problem is effected on the magnitudes in space, but the *actual* one by means of

---

* See Chap. V., "Cabinet Projection," in Div. IV. of my "ELEMENTARY PROJECTION DRAWING."

their projections, the two parts of the solution are appropriately entitled, "*in space*," and "*in projection*."

10. *Kinds of Problems*.—Problems naturally and conveniently fall under the successive heads of *projections*, meaning the representation of separate magnitudes; *tangencies*, relating to lines or surfaces which touch each other; *intersections*, relating to the cutting or penetration of one magnitude by another; *development*, which means the exhibition of a line, angle, figure, or curved surface, in its true size or form.

### Fundamental Problems.

11. *Fundamental problems* are those which relate to the immediate representation, without construction of required parts from given parts, of the most elementary magnitudes, the point, straight line, and plane; which, being such, are auxiliary to the solution of all other problems.

The importance of thorough familiarity with these problems as a means of readily proceeding with subsequent ones, leads to their separate formal treatment here.

### A—Projections.

PROBLEM I.—*To make the projections of points in various positions.*

*In Space.*—1°. The planes of projection in their real position (5, 7) being as represented pictorially (9 *second*) in Pl. I., Fig. 1, let $P_1$ represent a point in space, in the *first angle* (7), as signified by the subscript figure. Then by (6, 1°) $P_1 p_1$, perpendicular to H, represents the direction of vision in looking at H, and $p_1$, its intersection with H, is the representation of $P_1$ upon H as seen in the direction $P_1 p_1$. Likewise $P_1 p_1'$ being perpendicular to V, $p_1'$ represents $P_1$ upon V, as seen in the direction $P_1 p_1'$.

That is, $P_1 p_1$ and $P_1 p_1'$ are the *projecting lines* (5) of the point $P_1$, and $p_1$ and $p_1'$, respectively, its *horizontal projection* and its *vertical projection*. That is, the separate projections take their names from the plane on which they are found.

2°. In like manner, $P_2$ is in the *second angle*, and its horizontal projection $p_2$, and vertical projection $p_2'$, are on the *back* part of the horizontal, and the *upper* part of the vertical plane. $P_3$ is in the *third angle*, and its respective horizontal and vertical projections, $p_3$ and $p_3'$, are on the back part of the horizontal, and lower part of the vertical plane. Finally, $P_4$ is in the *fourth angle*, and its respective projections, horizontal and vertical, $p_4$ and $p_4'$ are on the front part of the horizontal, and lower part of the vertical plane.

*In Projection—First, pictorially.*—The planes **H** and **V**, being made to coincide as described in (8) Pl. I., Fig. 1, will be transformed as shown in Pl. I., Fig. 2, where this coincidence is intelligible by the lettering. Then making $mp_1$ and $mp_1'$ equal in both figures, and $mp_1'$ on $p_1m$ produced in Fig. 2, since it remains perpendicular to the axis $GL$ during revolution (8), $p_1$ and $p_1'$ will represent the projections $p_1,p_1'$ on Fig. 1, after **H** and **V** have been made to coincide.

Likewise, $p_2$ and $p_2'$ of Fig. 1 will appear as at $p_2$ and $p_2'$ of Fig. 2 on the same side of the ground line and beyond it. Then $p_3$ and $p_3'$ of Fig. 1 will appear as at $p_3$ and $p_3'$, Fig. 2, just the reverse of the projections of a point in the first angle, and $p_4$ and $p_4'$ of Fig. 1, at $p_4$ and $p_4'$, Fig. 2, on the same side of the ground line, but in front of it.

*Second, actually.*—The ideal planes of projection being of unlimited extent, are no longer shown in Fig. 3, as if limited as in a model, or in the picture of one, Fig. 1. $GL$ being the ground line, the planes **H** and **V** are represented as described in (8), and $a$ and $a'$ are the two projections, horizontal and vertical, respectively, of a point $A$ in space in the *first angle*, and which is at the height $a'm$ above **H**, and at the distance $am$ in front of **V**; as illustrated in Fig. 1, where $p_1'm = Pp_1$, and $p_1m = Pp_1'$.

Likewise $b$ and $b'$ are the projections, on **H** and **V** respectively, of a point $B$ in space in the *second angle*, at a distance from these planes, respectively, equal to $b'n$ and $nb$. Briefly, in like manner, $c$ and $c'$ are the projections on **H** and **V** of a point $C$ in the *third angle;* and $d$ and $d'$ are likewise those of a point $D$ in the *fourth angle.*

But a point may be in a plane of projection, whence (5) the point itself will coincide with its projection on the plane in which it lies, and its other projection will be on the ground line. See $E$, $e$ and $e'$, Fig. 1, representing a point $E$ in the front part of the horizontal plane. Thus, $e$ and $e'$, Fig. 3, represent such a point; $f$ and $f'$ are the projections of a point $F$ in the upper part of the vertical plane, and hence coinciding with $f'$; $g$ and $g'$ are those of a point $G$ in the back part of the horizontal plane; and $h$ and $h'$ those of a point in the lower part of the vertical plane; while $k$ and $k'$ are those of a point in the ground line $GL$.

EXAMPLES.—1°. Make a point in the first angle nearer to **H** than to **V**, and one in the third angle nearer to **V** than to **H**.

Ex. 2°. Letter this point ——⋮—— first, as in the first angle, and, second, as in the third angle, and in each case tell its distance both from **H** and **V**.

Ex. 3°. Make a point in the second angle nearer to **V** than to **H**; also letter this point ——⋮—— as nearer to **H** than to **V**, and tell which angle it is in.

12. *Determination of points.*—We now see from Problem. I:

1°. That the (orthographic) projection of a point upon a plane is where the perpendicular from that point to the plane meets the plane (5).

2°. That the two projections of the same point are necessarily on the same perpendicular to the ground line (8).

3°. That the perpendicular distance of a point in space from either plane of projection, is equal to the distance from the ground line to its projection upon the other plane. Thus, $P_1 p_1' = p_1 m$; $P_2 p_2 = p_2' n$, etc.

13. *Notation of Points—First, literal.*—It is evident, by comparing $a$ and $a'$, Pl. I., Fig. 3, with $c$ and $c'$, or $f$ and $f'$ with $g$ and $g'$, for example, that the position of a point can-

not be determined from its projections on Fig. 3 as it can on Fig. 1, without the aid of some uniform system of notation. The system usually adopted is the one here shown, viz., to give the unaccented letter to the horizontal projection, and the accented letter to the vertical projection, using additional accents, $a''$, $a'''$, etc., for other projections of the same point.

*Second, verbal.*—Since the projections of a point show (12, 3°) its distance from each plane of projection, and since a perpendicular to each plane at the projection of a point on that plane would meet at the point itself, as $p$, $P_1$ and $p_1'$ $P_1$ meet at $P_1$, Fig. 1, *a point is completely determined by its projections.* Hence, a point in space is considered as named by naming its projections. Thus the point $aa'$ means, the point $A$ in space whose projections are $a$ and $a'$.

A like rule applies to the projections of all magnitudes.

PROBLEM II.—*To draw the projections of lines in various positions in space.*

*In Space.*—A line may have a great variety of particular positions relative to the planes of projection, as perpendicular to either, parallel to either one only, or parallel to both. These are *elementary positions*, whose obvious projections are shown in Fig. 6. Then follow various *positions oblique to both planes*, with which it is important to become very familiar, and to be able to distinguish by the *angle which they cross*, and by the *direction in which they cross it*, as from right to left, or the reverse.

As a straight line is determined by any two of its points, its projections will be determined by the projections of those points. A particular case is that in which these points are the intersections of the line with the planes of projection; points which are called its *traces.*

Thus, in Pl. I., Fig. 4, the line $AD$ in space is determined by the two points $A$ and $D$—that is, by the points $aa'$ and $dd'$. Its projections are, therefore, $ad$ and $a'd'$, and it is briefly thereby designated (13) as the line $ad—a'd'$. If produced in the direction from $A$ to $D$, it would evidently pierce the front part of **H**, and if produced in the direction from $D$ to $A$, it would pierce the upper part of **V**. It therefore *crosses the first angle* and enters the *second* and *fourth* angles. $AD$ ($ad—a'd'$) also crosses the first angle from right to left as it descends from $A$ to $D$. The line $MN$, that is $mn=m'n'$, represents a line which is determined by its traces $m$ and $n'$.

The limited line $CO$ ($co—c'o'$) is in the third angle, and $Oo'$ being less than $Oo$, the line evidently pierces first the lower part of **V**, thence crosses the fourth angle from left to right, and next pierces the front part of **H**, where it enters the first angle.

In like manner, other positions can be represented.

*In Projection.*—Fig. 5 shows, pictorially, the projections alone, $ad—a'd'$, and $co—c'o'$, of the lines $AD$ and $CO$ in Fig. 4.

Fig. 6 shows the projections of the elementary and other positions. $a—a'b'$ is a vertical line, at a distance in front of **V** equal to $as$, and piercing **H** at the point $a$, whose vertical projection is $s$. The line $cd—c'$ is perpendicular to **V**. The line $ef—c'f'$ is parallel to both **H** and **V**, and hence to the ground line. $gh—g'h'$ is parallel to **H** only, as is indicated by drawing $g'h'$ parallel to the ground line (12, 3°). Its horizontal projection evidently shows its true length. $mn—m'n'$ is parallel to **V**, as shown by the direction of $mn$, parallel to the ground line, while $m'n'$ shows the true length of the line. The line $op—o'p'$ crosses the *first* angle *downwards from left to right;* while $qr—q'r'$ evidently meets **H** first, and thence crosses the *fourth* angle *downwards from left to right.*

Referring again to Fig. 4, it is evident that, by assuming the traces of a line, we can determine at pleasure which angle a line shall cross, and in what manner.

EXAMPLES.—1°. Assume successively the traces of lines that shall cross the first, second, third, and fourth angles respectively; first, from *right to left*, *as the line descends;* second, in the reverse direction, and draw the projections of the lines determined by such traces.

Ex. 2°. Draw lines in the *second* and *fourth* angles, as in Pl. I., Fig. 4, and their projections, as in Figs. 5 and 6.

Ex. 3°. Draw lines crossing the *first* angle at a large angle with **H** and a very small angle with **V**; also at a large angle with **V** and a small one with **H**; also like positions in each of the other angles.

### B—Tangencies.

14. Conceive a straight line tangent to a sphere at a point. Let the centre of the sphere recede from this tangent along the radius of the point of contact till it reaches an infinite distance from the tangent. The surface of the sphere will then become a plane containing the given line; which may still be said to be tangent to the plane, considered as an infinite sphere. We therefore have here the following problem:

**PROBLEM III.**—*To draw the projections of lines having various positions in the planes of projection.*

*In Space.*—By (12), if all the points of a line, and hence the line itself, lie in a plane of projection, the projection of the line upon the other plane of projection will be in the ground line. Such a line may be **parallel, perpendicular,** or **oblique** to the ground line.

*In Projection.*—Pl. I., Fig. 7. The line $ab$—$a'b'$ is in the front part of **H**, $kl$—$k'l'$ is in the upper part of **V**, and both are *parallel* to the ground line; $c$—$c'd'$ is in the upper part of **V**, and $ef$—$e'$ is in the back part of **H**, and both are *perpendicular* to the ground line. The three remaining cases are *oblique* to the ground line; $gh$—$g'h'$ in the front part of **H**; $mn$—$m'n'$ in the back part of **H**; $pq$—$pq''$ in the lower part of **V**.

EXAMPLE.—Draw such other parallels and perpendiculars to the ground line as are not here represented.

## C—Intersections.

**PROBLEM IV.**—*To represent planes in various positions.*

*In Space.*—Planes, other than those of projection, being, like those, supposed to be of indefinite extent, cannot, like limited magnitudes, be well represented by projections, since these would simply be in most cases the entire surface of both H and V. Thus, if *KMN* (Pl. I., Fig. 8) be a portion of an unlimited plane, its projections would be the whole surface of H and V. If, however, it has a fixed position, it will intersect H and V in fixed lines which will indicate its position. These lines *PQP″* on H, and *P′QP‴* on V, are called respectively the *horizontal* and *vertical traces of the plane*. Other magnitudes of indefinite extent may be similarly indicated.

As a plane can cut a line in but one point, *the traces of a plane must meet the ground line at the same point*.

The positions of a plane may be most simply classified as:

1st. In the same direction as the ground line. } containing it. parallel to it.

2d. In a different direction from the ground line. } perpendicular to it. oblique to it.

In Fig. 8, *RS* and *R′S′* being parallel to the ground line, represent the traces of a plane in like position. If *A′B′*, for example, on V, and parallel to *GL*, were the only trace of a plane, the latter would be parallel to H. Again, if *CD*, parallel to *GL*, and on H, were the only trace of a plane, the latter would be parallel to V.

*In Projection.*—Fig. 9 shows, pictorially, the traces *PQP″* and *P′QP‴* of the plane *PQP′*, Fig. 8, after H and V have been made to coincide.

In Fig. 10, *AB* is the single trace, coinciding with the ground line *GL*, of a plane containing *GL*. Such a plane is not determined unless by some further condition, as that it also contains a given point, or is parallel to a given line, or makes a given angle with H or V. *C′D′*, parallel to *GL*, is the vertical trace of a horizontal plane. *EF*, parallel to *GL*, is the horizontal trace of a plane parallel to **V**. *H* and *H'*, both parallel to *GL*, are the traces of a plane which is parallel to the ground line, but oblique to **H** and **V**, and crosses the *first* angle, as *RS—R'S'*, Fig. 8, crosses the third.

Again: *IJI'* is a plane, perpendicular to the ground line, and hence to both **H** and **V**. Such a plane is often called a *profile plane*. *KMK'* is a plane perpendicular to **V** only and *NON'* is perpendicular to **H** only. The last two positions are of especially frequent use, and may be fixed in mind by noting that *when a plane is perpendicular to either plane of projection, its trace on the other plane of projection is perpendicular to the ground line.*

*PQP'* is oblique to both **H** and **V** and to the ground line as at *PQP'*, Fig. 8—that is, in such a way as to make *acute angles* with both **H** and **V** in the first angle, on the *same side of itself;* while *RSR'* is a plane of which the part in the first angle is situated as *P''QP'* is in Fig. 4—that is, so as to make an *obtuse* angle with **V** and an *acute* angle with **H** on the same side of itself—the right-hand side in Fig. 10. The angles made by the *traces*, on **V** and **H** respectively, *with the ground line*, on the same side, are just the reverse of these relative to the same planes.

15. The projecting lines of all points of a given line, evidently form a pair of *projecting planes* of that line, one of them perpendicular to **H**, and the other to **V**. It is thus evident that the two *projections* as well as the two *traces* of a line determine the line, since the projecting planes being erected on the projections could only intersect in the one line represented by those projections.

An exception to this principle occurs when both projections of the line are perpendicular to the ground line, for then the two projecting planes of the line coincide. The line must then be determined by two of its points.

It is now clear that the plane, as $p_,mp_,'$, Pl. I., Fig. 1, of the two projecting lines of any point is perpendicular to the ground line; hence the projections of these lines, being the traces of this plane, *meet the ground line at the same point*. Hence, after the revolution of **V** to coincide with **H**, these traces will form a perpendicular to the ground line, and joining the two projections, $p_1$ and $p_1'$, of the same point $P_1$ (12, 1°).

16. *Notation of Lines and Planes—First, literal.*—The projections of a line, like those of a point (13), are distinguished by accents attached to the letters which denote them, and a line, like a point, is named in naming its projections. Thus $ab$—$a'b'$ means the line $AB$ in space whose horizontal and vertical projections are respectively $ab$ and $a'b'$. A line may be, but not often is, designated by its traces.

A plane is usually designated by lettering and naming its *traces*, as indicated in Pl. I., Fig. 10.

*Second, graphical.*—Invisible and auxiliary *lines* are dotted. The planes of projection being supposed to be opaque, nothing is visible out of the first angle (7). The projections of lines are inked, not according to their own visibility, but according to that of the line itself which they represent.

The traces of invisible and auxiliary *planes* are made in broken and dotted lines, consisting of dashes alternating with dots.

Sometimes *invisible* lines are distinguished from *auxiliary* ones, otherwise called *construction lines*, by inking the projections of the former with true dots, *point dots*, and the latter by *dash dots*, or very short dashes.

Construction lines may also be *colored*, in which case they need not be dotted.

17. *Verbal notation.*—The terms *horizontal* and *vertical* briefly and clearly express the almost constantly recurring positions, *parallel to the horizontal plane of projection*, and *perpendicular to the horizontal plane of projection*. As the positions, parallel to the vertical plane; perpendicular to the vertical plane, or to the ground line; and parallel to the ground line, all occur about as frequently as the two former ones, similarly brief terms to denote these positions,

like profile plane (Prob. IV.) for the *third*, would greatly abridge speech in the study of descriptive geometry. The following are therefore suggested for optional use in the other cases: *Co-parallel*, for parallel to **V**, the two planes **H** and **V**, and their contents, being complementary to each other, and parallel to **V** being the same thing relative to **V** that horizontal is to **H**; then *perpendicular*, for perpendicular to **V**, which is merely extending the established usage of perspective to the present case; finally, *bi-parallel*, as clearly expressive of parallel to the ground line, that position being also parallel to both **H** and **V**.

### D—Development.

18. A fundamental operation in many practical problems of development is the revolution of a point, as shown in the following problem; where the result, being confined to the simplest positions relative to the planes of projection, is immediately shown without construction (10).

In every revolution about an axis, we have to remember:

1°. That every point in the axis remains fixed.

2°. That every other point moves in a circle whose plane is perpendicular to the axis, whose centre is in the axis, and whose radius is the perpendicular distance of the point from the axis.

3°. That the revolving points preserve their relative position.

PROBLEM V.—*To revolve a point so as to immediately show the form and extent of its path.*

*In Space.*—The point must, to fulfil the requirement, revolve in a circle parallel to a plane of projection; hence, the axis about which it revolves must be perpendicular to the same plane.

*In Projection.*—Pl. II., Figs. 11, 12. In Fig. 11, let the vertical straight line $C$—$C'C''$ be a fixed axis about which a point, whose horizontal projection is $a$, revolves. By (18) this point must therefore describe a *horizontal* circle, whose *horizontal projection would be an equal circle,* and all points of this circle being at equal heights above **H**, must therefore (12, 3°) be vertically projected in a *straight line parallel to the ground line.* Then, if the given point were in **H**, at $aa'$, 1 1', 2 2', 3 3', etc., would be projections of its successive positions after revolving through the arcs $a1—a'1'$, $a2—a'2'$, etc. But if the point were in space, as at $aa''$, then 1 1'', 2 2'', 3 3'', etc., would be its successive positions after like revolutions in a plane at the height $a'a''$ above **H**.

Fig. 12 shows various positions of a point, $aa'$, in the vertical plane of projection, and of a point, $a''a'$, at the distance, $aa''$, in front of **V**, in the revolution of these points about an axis, $C''C—C'$, which is perpendicular to **V**.

EXAMPLES.—1°. In Fig. 11 let the given revolving point be taken successively in each of the other angles besides the first.

Ex. 2°. Make the same changes in the position of the given point in Fig. 12.

## Classification of Surfaces and Lines.

19. *The generation of a magnitude* is its formation by the successive positions of some simpler moving magnitude. Thus the motion of a *point* produces a *line* of some form; a *moving line* describes a *surface* of some kind. The moving magnitude is called a *generatrix;* any fixed magnitude employed to guide its motion is called a *directrix*, if it be a *point* or *line*, a *director* if it be a *surface.* The various positions of a linear generatrix are called *elements* of the surface which they form, and immediately successive points or elements are called *consecutive.*

20. *Revolution and Transposition.*—For the purposes of descriptive geometry, the main divisions of surfaces are conveniently made to depend on the *form* or the *motion* of the generatrix.

When the *motion* of any generatrix whatever is one of revolution around a fixed axis, every one of its points describes a circle centred in and perpendicular to that axis (18). All surfaces so formed are thence called SURFACES OF

REVOLUTION, and possess the important property that their plane sections, perpendicular to the axis, are *circles*.

When the motion of the generatrix is not one of revolution, it is said to be one of *transposition*. Surfaces thus generated are thence called *surfaces of transposition*, and they either possess no line which could be called an axis, or none containing the centres of a series of circular sections perpendicular to it.

21. *Ruled and double curved surfaces.*—Again: all surfaces, whether of revolution or of transposition, which can be generated by the motion of a *straight* line, are called *ruled surfaces*, since such lines can be ruled upon them.

All surfaces which must be generated by a *curved* line are called *double curved surfaces*, since, as any surface has but two dimensions, and these surfaces are straight in no direction, *they are curved in the directions of both of their two dimensions*, from whatever point, and on whatever pair of tangent directions at right angles to each other these dimensions are estimated.

22. *Principal subdivisions.*—RULED SURFACES are of two kinds, *plane* and *single-curved*. A *plane surface*, considered as a surface of revolution, is generated by the revolution of a straight line, either about an axis which is *perpendicular* to it, or which is *parallel to it and at an infinite distance* from it.

A *single curved surface* of revolution is generated by the revolution of a straight line about an axis, which it *intersects obliquely*, as in the *cone of revolution* which is that of elementary geometry; or to which it is *parallel*, as in the *cylinder of revolution*, which is also that of elementary geometry; or, finally, with which it is *not in the same plane*.

In the two former cases, consecutive elements (19) are evidently in the same plane. Such surfaces being cut along any element, may be unfolded or developed upon a plane, and are therefore called *developable surfaces*.

In the last case, the surface formed does not possess this property, and is called a *warped surface*.

23. The foregoing are all the possible forms of *ruled surfaces of revolution*, since they exhaust all the possible relative positions of the straight lines, one of which is the *generatrix* and the other the *fixed axis*.

The plane is evidently that particular case of the cone in which the generatrix is *perpendicular* to the axis, and that one of the cylinder in which the generatrix is *infinitely distant* from the axis.

24. DOUBLE CURVED SURFACES, alike of revolution and of transposition, are either *doubly convex*, as a *sphere*, on which any two great circles at right angles to each other are both convex towards surrounding space, or both concave on the side towards their common diameter considered as an axis; or are *concavo-convex*, as the *surface of a bell* in which the rim and other circular sections are convex towards surrounding space while the bell-shaped sections are concave towards it.

25. *Lines.*—These, however imagined as independent of surfaces, always arise practically as the lines of contact or of intersection of surfaces. They are classed in descriptive geometry as *straight* or *curved*, and curved lines as of two kinds, *curves of single curvature*, or such as lie in one plane, as a circle; and *curves of double curvature*, which do not.

A *straight line* is generated by a point moving parallel to a *fixed straight line* (19), or directly towards a *fixed point*.

# DESCRIPTIVE GEOMETRY.

## BOOK I.—SURFACES OF REVOLUTION.

### PART I.—RULED SURFACES.

#### CHAPTER I.

GENERAL PROBLEMS OF THE POINT, LINE, AND PLANE.

26. In all the operations of descriptive geometry, lines and planes are supposed to be of indefinite extent, so that the *projections* of the former and the *traces* of the latter can always be produced indefinitely in either direction.

#### A—Projections.

PROBLEM VI.—*Having given two projections of a point and a line, to find their projections on any new planes of projection.*

*In Space.*—By (12, 3°) the heights of the projections of any fixed point upon any number of vertical planes, and measured from their respective ground lines, will be equal, being all equal to the height of the given point above the horizontal plane **H**. Likewise, the projections of one point, upon different planes perpendicular to the vertical plane **V**, will be at equal distances from the traces of these planes upon **V**.

*In Projection.*—Pl. II., Fig. 13. Let $AB$—$A'B'$ be a given line, and let its projection upon the vertical plane at $MN$ be found. By (12) draw $AA''$ and $BB''$ perpendicular to $MN$, and make $r_1A'' = rA'$ and $s_1B'' = sB'$, and $A''B''$ will be the projection of $AB$—$A'B'$ upon $NMN'$ after the revolution of this plane to coincide with **H**.

Again, let $aa'$ be a given point, and let its projection on a plane, $XYX'$, perpendicular to **V**, be found. Then, drawing $a'a''$ perpendicular to $X'Y$, and making $p_1a'' = pa$, we find $a''$, the required projection of $aa'$ on the new plane $XYX'$.

EXAMPLES.—1° Let $aa'$ be in any other angle than the *first*, and find its projection also on a new vertical plane.

Ex. 2°. Let $AB$—$A'B'$ be in any other *position* (Prob. II.), or in any other *angle* than the *first*.

Ex. 3°. Having given the traces of a line to find its projections.

Ex. 4°. Through a given point, draw lines parallel to any two given lines.

## B—Tangencies.

PROBLEM VII.—*To draw a straight line in a plane which is given by its traces, and a plane to contain a line which is given by its traces.*

*In Space.*—We have seen (Prob. IV.) that a plane is determined by its traces. It is also determined by any two intersecting or parallel lines, $A$ and $B$; for let it at first contain only $A$, then by revolving about $A$ as an axis, it must at some one and only one position contain $B$ also, when it will be determined.

If a line contains two points of a plane, it lies wholly in the plane; hence *a line which intersects both of any two lines in a plane lies in that plane.*

Again: any line which intersects a given line will determine a plane containing the latter, and if the traces of the given line are known, lines from them to any one point (Prob. IV.) on the ground line will be the traces of a plane containing the line.

*In Projection.*—Pl. II., Figs. 14, 15. In both, let $PQP'$ be the given plane. Then any line, $ab$—$a'b'$, from any point, $aa'$, on $PQ$ to any point, $bb'$, on $P'Q$, will be in $PQP'$.

Or, if the plane be given, Fig. 14, by two intersecting lines, $ab$—$a'b'$ and $cd$—$c'd'$, then any line which intersects both of these—including the particular case of intersecting either and being parallel to the other—will lie in the plane of the given lines.

Again, any line, $cd$—$c'd'$, drawn (12, 2°) so as to intersect a given line, $ab$—$a'b'$, as at $nn'$, will determine a plane through $ab$—$a'b'$.

Or if, as in the figures, the traces $a$ and $b'$ are given, any point, $Q$, of the ground line, joined with these traces, will give the traces of a plane containing $ab$—$a'b'$.

EXAMPLES.—1°. Let the plane have any of the positions shown in Prob. IV., with any variations of the *kinds* of positions shown in Figs. 14 and 15.

Ex. 2°. Let the traces of either the given or the required line be so chosen as to make these lines cross any of the four angles.

Ex. 3°. Let each trace in succession of the line be at an infinite distance from $Q$. [The line will then be *parallel* to that *trace* of the *plane* which contains this trace of the *line*.]

Ex. 4°. Draw two lines which shall be in the same vertical plane.

Ex. 5°. Draw two lines which shall both be in a plane perpendicular to V.

Ex. 6°. Having given one projection of a line in a given plane, find its other projection.

## C—Intersections.

PROBLEM VIII.—*To find the traces of a line given by its projections.*

*In Space.*—The required traces, being points of a given line, the projections of each of them will be on the projections of the line. Each trace, being also in a plane of projection, will coincide with its projection on that plane, while its other projection will be on the ground line. Combining these two principles, we have the following:

Rule. *To find the trace of a line on either plane of projection:*

*First.* Note where the projection of the line on the *other* plane of projection intersects the ground line.

*Second.* Thence draw a perpendicular to the ground line, and its intersection with the projection on the proposed plane will be the required trace.

Thus, Pl. II., Fig. 16, a picture model of the problem, shows a line, $AB$, in space, piercing the plane **V** at $y'$, thence crossing the *second* angle, and piercing the back part of **H** at $x$. But, according to the rule, $x'$, the vertical projection of $x$ is the intersection of $a'b'$ with the ground line $GL$, and it must be found first, when, as in practice, the line itself, $AB$, is not shown. Likewise, $y$, found before $y'$, is the intersection of $ab$ and $GL$.

*In Projection.*—Pl. II., Figs. 17, 18. In both figures, $x$, the horizontal trace of the given line, $ab$—$a'b'$, and $y'$, its vertical trace, are found just as above directed and illustrated, as may be seen by inspection.

Examples.—1°. Find the traces of a line crossing the *third* angle, and of one crossing the *fourth* angle. [Figs. 17 and 18 inverted, without change of notation, will represent these cases.]

Ex. 2°. Take the given portion of the line in each of the four angles successively, and then find its traces.

Ex. 3°. Let the line, determined by two given points, be in a plane perpendicular to the ground line. [In this case, first project the line upon a new plane of projection, as in Prob. VI.]

Ex. 4°. Assume pairs of traces in many different positions in any part of both planes of projection, and then construct the projections of the lines to which these traces belong.

27. Difficulties at the outset, in the solution of problems, being often caused by oversight of, or uncertainty about very elementary details, the following are, for convenience, added or repeated here:

1°. If a point is on a line, its projections will be on the projections of that line.

2°. If two planes are parallel, their traces will be parallel.

3°. If lines are *parallel*, their projections on the same

plane will be parallel, for their projecting planes (15) are parallel, and hence the traces of these planes, which are the projections of the lines, will be parallel also.

4°. If a line lies in a plane, its traces will lie in those of the plane; or if a plane contains a line, its traces will contain those of the line (Prob. VII.).

5°. A *horizontal* line in a plane is parallel to the *horizontal trace* of the plane, and its *vertical projection* is therefore *parallel to the ground line*. Likewise, and with a slight change in the form of statement, a line in a plane and *parallel to its vertical trace* is parallel to **V**, and hence its *horizontal projection is parallel to the ground line*.

6°. Two straight lines can intersect only at one point, hence their projections will intersect at points on the same perpendicular to the ground line (12, 2°). If the projections intersect otherwise, the lines themselves do not intersect. See Pl. II., Fig. 14, where $ab$—$a'b'$ and $cd$—$c'd'$, being in the same plane, $PQP'$, intersect at $nn'$.

7°. If any point, line, or figure be in one plane of projection, its projection upon the other plane will be in the ground line (Prob. I.). Hence,

8°. If a plane be perpendicular to a plane of projection, its trace on that plane will contain the projections on that plane, of every thing in the given plane.

9°. If a plane contains two lines, it will contain any line which intersects both of them.

THEOREM I.—*The projection of a right angle will be a right angle when at least one of its sides is parallel to the plane of projection.*

For if *both* sides are thus parallel, their projecting planes will be perpendicular to each other; hence their traces, which are the projections of the sides of the angle, will be so also. But either of these sides, $A$, being revolved about the other one, $B$, as an axis, it will evidently generate a plane (23) perpendicular to that other side, $B$, and hence to any plane containing $B$. That is, the projecting planes of the two sides, and hence the directions of the projections of *A* and *B*, remain unchanged.

THEOREM II.—*If a line is perpendicular to a plane, its projections will be perpendicular to the traces of the plane.*

Pl. II., Fig. 19. Let *AB* be perpendicular to the plane *PQP′* at *B*. It will therefore be perpendicular to *every* line drawn through *B*, and in the plane *PQP′*, and hence to a parallel, *R* (not shown) to *PQ*, through *B*. But the projecting planes of *AB* and *R* are (Theor. I.) perpendicular to each other, and the horizontal projection of *R* is (27, 3°) parallel to *PQ*. Hence the horizontal projection *ab* of *AB* is perpendicular to *PQ*. Likewise *a′b′* is perpendicular to *P′Q*.

*Otherwise*, either projecting plane, as *ABab*, of the line *AB*, is perpendicular to a plane of projection, as H, and to *PQP′* because containing *AB*, a perpendicular to the latter plane. Hence it is perpendicular to the intersection, as *PQ*, of H and *PQP′*. Hence, for example, the traces of *ABab*, on H and on *PQP′*, are both perpendicular to *PQ*. That is, *ab* is perpendicular to *PQ*, and the like is true of *a′b′* and *P′Q*.

PROBLEM IX.—*To construct the traces of a plane under various frequently occurring conditions.*

*In Space.*—Many particular problems are here included in this comprehensive title; partly since the plane, as the simplest of surfaces, is auxiliary to the solution of many problems where its construction is required under various conditions; but especially since the main feature of the solution in all cases is the same, viz., the *determination of two lines which shall be in the plane*, and whose traces will therefore (Prob. VII.) determine those of the plane.

*The principal cases* and their determining lines are as follows: To construct a plane:

1°. *Through three given points*, the determining lines being then *any two of the three lines connecting these points.*

2°. *Through a given point and straight line.* Here the lines are *the given line and one from any point of it to the given point*, usually a *parallel to the given line, through the given point.*

3°. *Through two given parallel or intersecting lines;* when these, or, in case of their inconvenient location, any lines intersecting both of them (Prob. VII., 27, 9°), will be the determining lines of the plane.

4°. *Through one given line and parallel to another;* when the plane will be determined by the former line, together with a line through any point of it and parallel to the second line.

5°. *Through a given point and perpendicular to a given line;* when each of the determining lines will contain the given point and will be parallel to a trace of the plane. Hence (Theor. II.) the projection of either of these lines on the plane of projection containing such trace will be perpendicular to the projection of the given line on the same plane (27—3°, 5°).

6°. *Through a given point and parallel to any two given lines.* The determining lines will pass through the given point and be parallel to the given lines.

7°. *Through a given point and parallel to a given plane;* when the plane will be determined by lines through the given point and parallel to any two lines taken in the given plane by Prob. VII.

*In Projection.*—The construction of cases 2°, 4°, and 5° will sufficiently illustrate all the rest, on account of the similarity of all, as stated.

*Case* 2°.—Pl. II., Fig. 20. This illustrates the particular case, frequently occurring, where the given line is one trace of the plane, and the given point is assumed on some known line of the plane. Then let $Pc$ be the given trace, and $aa'$, on the known line $cd$—$c'd'$, the given point. Through $aa'$ draw $ab$—$a'b'$ parallel to $Pc$. Its vertical trace, $b'$, together with $d'$, the vertical trace of $cd$—$c'd'$, will determine the required vertical trace.

In reference to frequent variations occurring in practice, note that if $d'$ were inaccessible, the vertical trace $P'Q$ would join $b'$ with the intersection $Q$ of the horizontal trace with the ground line, but if $Q$ and $d'$ were both inaccessible, we should take an auxiliary line through another point of $cd$—$c'd'$ and parallel either to **H** or to **V**.

*Case* 4°.—Pl. II., Fig. 21. Let $ab$—$a'b'$ be the line which the required plane is to *contain*, and let $cd$—$c'd'$ be the one to which it is to be *parallel*. Assume any point, $pp'$, on $ab$—$a'b'$, and through it draw $c_1d_1$—$c_1'd_1'$ parallel to $cd$—$c'd'$. Then the horizontal traces, $a$ and $d_1$, of these lines, being joined, give $PQ$ the horizontal trace of the required plane. Likewise the vertical trace $P'Q$ joins $b'$ and $c_1'$, the vertical traces of these lines. Two points being sufficient for each trace, $Q$, being common to both traces, serves as a test of the mechanical accuracy of the construction.

*Case* 5°.—Pl. III., Fig. 22. Let $ab$—$a'b'$ be the line to which a plane through the given point $pp'$ is to be perpendicular. Then $pc$ being perpendicular to $ab$ and $p'c'$ parallel to $GL$, the line $pc$—$p'c'$ is a parallel to **H** in the required plane; hence its vertical trace, $c'$, is a point of the vertical trace $P'Q$, perpendicular to $a'b'$, of the required plane. Likewise, $pd$ being parallel to $GL$, and $p'd'$ perpendicular to $a'b'$, $pd$—$p'd'$ is a parallel to **V** in the required plane; hence, finding its horizontal trace $d$, draw $PQ$ through $d$ and perpendicular to $ab$, for the horizontal trace of the plane.

*Remark.*—The line from the *given point* to the intersection of the plane with the *given line* is the *perpendicular* from the given point to the given line.

EXAMPLES.—1°. Construct *case* 1° with variations by taking the three given points in one angle, or in two or three different angles.

Ex. 2°. Construct *case* 1° when one of the determining lines joining two of the points is parallel to a plane of projection.

Ex. 3°. Construct *case* 1° when the same line is parallel to the ground line.

Ex. 4°. Construct *case* 2° in its general form, the given line and point having any position in the same or different angles.

Ex. 5°. Construct *case* 3° when the traces of either or both of the given lines are inaccessible.

Ex. 6°. Construct *case* 3° when either or both of the given lines are parallel to the ground line.

Ex. 7°. Construct *case* 3° when the given lines intersect on the ground line.

28 DESCRIPTIVE GEOMETRY.

Ex. 8°. Construct *case* 4° when the given lines, which in Fig. 21 both cross the *first* angle, shall cross different angles.

Ex. 9°. Construct *case* 5° when the given line crosses either the second or the fourth angle, and when the same line is either parallel or perpendicular to a plane of projection.

Ex. 10°. Construct *case* 5° when the given line is in a *profile plane* (Prob. IV.).

Ex. 11°. Construct *case* 6°.

Ex. 12°. Construct *case* 7°.

Ex. 13°. Construct a plane parallel to a given plane and at a given perpendicular distance from it.

## PROBLEM X.—*To construct the intersection of two planes given by their traces.*

A line may be cut from each given plane by an auxiliary plane. The intersection of these lines will be a point of the line of intersection of the given planes. The planes of projection may generally be the auxiliary planes, and the intersection of the given planes then joins the intersection of their vertical traces with that of their horizontal traces.

Thus, Pl. III., Fig. 23, **H** and **V** being the planes of projection, $PQP'$ and $RSR'$ are two planes whose like traces intersect at $a$ and at $b'$. That is (27, 4°), *the intersections of the like traces of two planes are the traces of the intersection of the planes.* But one projection of each trace is on the ground line (27, 7°); hence $ab$ and $a'b'$ are the *projections* of the intersection $ab'$. Hence, when the planes **H** and **V** are the auxiliary planes, we have the

RULE.—*Project the intersection of the traces on each plane of projection upon the ground line, and join the points so found with the intersection of the traces on the other plane.*

*In Projection—The general case.*—Pl. III., Figs. 24, 25. In both, where like letters designate like points *similarly found*, though in *different positions* only, owing to the different positions of the given planes in the two cases, let $PQP'$ and $RSR'$ be the given planes; then $a$ and $b'$ are the *traces* of their line of intersection, whose projections are therefore $ab$ and $a'b'$.

*Particular cases—the traces of the intersection inaccessible.*

1°. *The planes parallel to the ground line.*—Pl. III., Fig. 26. The intersection of these planes, $MN$—$M'N'$ and $XY$—$X'Y'$, will also be parallel to the ground line. Thus, knowing its direction, one point will determine it. Take any *auxiliary* plane, as $NxX'$, which will conveniently intersect both of the given planes. By the *general case*, above, this plane cuts from $MN$—$M'N'$ the line $Nx$—$n'M'$, and from $XY$—$X'Y'$ the line $rx$—$r'X'$. These lines intersect at $t'$, whose horizontal projection is $t$, whence $ts$—$t's'$, parallel to the ground line, is the required intersection.

A like construction, with two auxiliary planes, would have served had the planes been *nearly parallel to the ground line*.

2°. *The planes nearly perpendicular to the ground line.*—In this case, the auxiliary planes are conveniently taken parallel to **H** or **V**. Thus, Pl. III., Fig. 27, let $PQP'$ and $RSR'$ be the given planes. Then a horizontal plane, $C'D'$, will cut the given planes in lines parallel to their horizontal traces. The vertical traces of these lines are by (27, 4°) $C'$ and $D'$, hence their projections are $ck$—$C'k'$ and $dk$—$D'k'$, where $ck$ and $dk$ are parallel to $RS$ and $PQ$, and they give the point, $kk'$, of the required intersection.

Likewise $AB$ is the horizontal trace of an auxiliary plane parallel to **V**, and cutting the given planes in lines parallel to their vertical traces, and beginning at $A$ and $B$. Hence, project $A$ and $B$ at $a'$ and $b'$, draw $a'h'$ and $b'h'$ parallel to $SR'$ and $QP'$, and meeting at $h'$, whence project $h'$ to $h$ on $AB$, and we have $hk$—$h'k'$ for the required intersection.

Had the planes been *very* nearly perpendicular to the ground line, we could have used, for *each* projection of the intersection, two auxiliary planes parallel to the ground line, and very nearly so to the plane containing the other projection of the intersection.

EXAMPLES.—1°. Construct the case just mentioned.

Ex. 2°. Take planes whose intersection shall cross the *third* angle and the *fourth* angle, and find the intersection in each case.

Ex. 3°. Let one of the planes be perpendicular either to **H** or to **V**, and treat it as a *new plane of projection* by revolving it into the plane to which it is perpendicular, thence showing a trace of the given plane on a new plane of projection.

Ex. 4°. Let the traces of the given planes on *one* of the planes of projection be parallel to each other.

Ex. 5°. Let both planes cut the ground line at the same point.

Ex. 6°. Let planes, both parallel to the ground line, be so taken as to intersect in the *fourth* angle.

Ex. 7°. Let one of two given planes be parallel to the ground line, and cross either the *second* or *fourth* angles, so that its two traces shall coincide on the paper.

Ex. 8°. Find the intersection of two planes each of which is given by two lines contained in it.

Ex. 9°. Find the *point* of intersection of *three* given planes.

PROBLEM XI.—*To construct the intersection of a line with a plane.*

*In Space.*—Pl. III., Fig. 28. In this picture model, **H** and **V** are the planes of projection, *PQP'* is any given plane, and *AD* a given line whose projections are *ad* and *a'd'*. *AD* evidently will meet *PQP'* by (27, 4°) in the trace upon *PQP'* of any plane containing *AD*. The projecting planes of *AD* being the most convenient, *Aann'* represents that which projects it upon **H**, and whose trace on *PQP'* is *mn'*, projected in *mn* and *m'n'* (Fig. 23). Hence, *AD* pierces *PQP'* on *mn'* at *B*, whose projections are *b'* and *b*, on *m'n'* and *mn*.

Likewise, taking the vertically projecting plane *Aa'h'h*, of *AD*, its trace on *PQP'* is *k'h*, projected in *hk* and *h'k'*. This meets *AD* at *B* as before. Hence the

RULE.—*Find the intersections of the projections of the given line, with the projections of the intersection of the given plane with either of the projecting planes of the given line.*

*In Projection—a. The general case.*—In Pl. III., Fig. 29, *PQP'* is the given plane (differently situated from *PQP'* in Fig. 28, to further illustrate, as in Figs. 24 and 25, the following of the *same operations* in different *positions* by the aid of like letters at like points), *ad—a'd'* is the given line, *ann'* is its horizontally projecting plane, intersecting the given plane (Prob. X.) in the line *mn—m'n'*, which, in turn, intersects

$ad$—$a'd'$ at $b'b$ (naming $b'$ first because it is found first), the required intersection of $ad$—$a'd'$ with $PQP'$.

Likewise, the vertically projecting plane, $d'h'h$, intersects $PQP'$ in the line whose traces are $h$ and $k'$, and whose projections are $hk$ and $h'k'$, giving $bb'$, as before, for the required intersection, only that $b$ is thus found before $b'$.

NOTE.—Problems VIII., X., and XI. have thus been emphasized by *pictorial illustration*, and by the expression of their solution in a *concise rule*, because they are the ones most frequently applied in the solution of all other problems. They should therefore be made thoroughly familiar by review and varied examples before proceeding further.

*b. Particular cases.*—1°. *The given line perpendicular either to* **H** *or* **V**. In this case, the projection of the entire line on the plane of projection to which it is perpendicular is a point (Prob. II.) which is therefore one projection of the required intersection of this line with the given plane. Hence this case is equivalent to the problem, often separately enunciated, "*Having one projection of a point of a given plane, to find the other projection of the same point.*"

In this case, *any* auxiliary plane containing the given line will be perpendicular to that plane of projection to which the given line is perpendicular, and will therefore be convenient.

Thus, let $PQP'$, Pl. IV., Fig. 30, be the given plane, and let $a$ be the horizontal projection of a vertical line, $a$—$a'b'$. Then *any* line, $MN$, drawn through $a$, will be the horizontal trace of a vertical plane, $MNM'$, whose trace, $M'm'$—$MN$, on $PQP'$ will contain the required intersection, $c'c$, of $a$—$a'b'$ with $PQP'$. In other words, $c'$ is the vertical projection of that point of the plane $PQP'$ whose horizontal projection is $c$.

Otherwise (and a little more convenient instrumentally), let $p$ be the horizontal projection of a vertical line whose intersection with $PQP'$ is to be found. Then $RSR'$ is a vertical plane containing $p$ and parallel to $PQ$, and hence cutting $PQP'$ in a horizontal line, $RS$—$R'r'$, which contains the intersection, $pp'$, of the vertical line, $p$—$p'q'$, with the plane $PQP'$.

$RS$ could equally well have been parallel to the ground line $GL$, when $R'r'$ would have been parallel to $QP'$.

The same lines here described, only drawn in a different order, would have served to find $p$ and $a$ had $p'$ and $c'$ been given.

EXAMPLES.—1°. Let the given line be parallel to a plane of projection.
Ex. 2°. Let it be parallel to the ground line.
Ex. 3°. Let the plane in Fig. 29 take the position of $PQP'$ in Fig. 30, the given line remaining the same.
Ex. 4°. Let the given line cross any one of the four angles.
Ex. 5°. Let the plane be parallel to the ground line.
Ex. 6°. Let the traces of the plane coincide on the paper.
Ex. 7°. Let the line be in a profile plane (Prob. IV.).

2. *The plane given by other lines than its traces.*—The solution is the same as in Fig. 29, except that we note the intersection of one or both auxiliary projecting planes of the given line with these determining lines of the plane, rather than with the traces of the plane, in order to find the trace of this projecting plane upon the given plane. Thus, Pl. IV., Fig. 31, let the given plane be that of the two intersecting lines $ab$—$a'b'$ and $cb$—$c'b'$; and let $RS$—$r's'$ be the line whose intersection with this plane is required. Then, as before, $RSR'$ is a projecting plane of the given line $RS$—$r's'$, and it intersects $ab$—$a'b'$ at $ee'$, and $cb$—$c'b'$ at $dd'$, giving $ed$—$e'd'$ as the trace of this projecting plane upon the given plane. Hence, $q'q$, the intersection of this trace with $RS$—$r's'$, is the required intersection of the line $RS$—$r's'$ with the plane $abc$—$a'b'c'$.

EXAMPLES.—1°. Let the point $bb'$ be in any other angle than the *first*.
Ex. 2°. Let the given line cross any other angle than the *first*.
Ex. 3°. Construct a line parallel to a given line, $A$, and intersecting two other given lines, $C$ and $D$, which are not in the same plane. [Pass a plane, by Prob. IX., *case* 4°, through $C$ and parallel to $A$, and, by Prob. XI., find where $D$ pierces this plane. A line from the latter point, parallel to $A$, will be the one required.]

NOTE.—Problems VIII. and XI. will enable the student to take up perspective constructions by the method of visual rays (5) according as **V** or some other plane is taken as the perspective plane ; also to find shadows on plane surfaces, if the given lines be supposed to be rays of light.

DESCRIPTIVE GEOMETRY. 33

## D—Development.

28. As a plane cannot be reduced to any simpler form by *transformation*, the only problems of development here to be considered are those of *transposition*, the object of which is to show a line, angle, or figure in its real size by bringing it into or parallel to a plane of projection. Such problems are *direct* when the developments are the results required by them, and *inverse* when from *auxiliary* developments required projections or other developments are found.

### General Methods of Solution.

29. *Three general methods of solution* are employed in descriptive geometry, which, as they are best illustrated under the present head of development, are stated here.

*First.* As evidently illustrated by any of the preceding problems, beginning with Pl. II., Fig. 20, the planes of projection and the given magnitudes are both fixed; and the required parts are found by operating upon the given ones by auxiliary lines and planes suitably chosen for the purpose.

*Second.* The planes of projection being fixed, the given magnitudes may be brought by one or more rotations, usually elementary rotations—that is, about an axis perpendicular either to H or V, as in Prob. V.— to such a simple position relative to the planes of projection, that the parts required there immediately appear without further construction. This is the *method of rotations*.

*Third.* The given magnitudes being fixed, their projections may be successively found on one or more new planes of projection as in Prob. VI., until by that means the same result is produced as by the method of rotations. This is the method of *change of planes of projection*.

A comparison of the following problems with these statements will sufficiently illustrate them.

PROBLEM XII.—*To find the true length of the line joining two given points.*

*In Space.*—A line may be shown in its true length by bringing it to a position *in or parallel to a plane of projection.*

This position may be obtained by revolving the line either about an axis, *perpendicular* to one plane of projection till it becomes parallel to the other one (29, *Second*), or about an axis, *in or parallel* to a plane of projection, till the line reaches a like position.

In the *former method*, the points of the line will move in arcs parallel to a plane of projection as in Prob. V. By the *latter method*, we employ *the principle* that the distance of a point in space from any line taken as an axis is the *hypothenuse* of a right-angled triangle, whose other two sides are the *projection of this distance upon a plane containing the axis*, and the projecting line of the point upon this plane.

*In Projection.*—Pl. IV., Figs. 32, 33. *First method:* the axis perpendicular to **H** or **V**. In Fig. 32, let $ab$—$a'b'$ be the given line, and let it be revolved about the vertical axis at $b$ till parallel to **V**. The point $aa'$ will then describe the *horizontal* arc $aa_1$—$a'a_1'$, having the point $bb''$ for its centre, and limited by $ba_1$—$b'a_1'$, parallel to **V**. Hence $ba_1$ is parallel to the ground line, and $a_1'b'$ is the true length of the line $ab$—$a'b'$.

Note that such an axis need not intersect $ab$—$a'b'$, but may be *any* perpendicular to **H** or **V**.

*Second method:* the axis in or parallel to **H** or **V**. In Fig. 33, let $ab$—$a'b'$ be a given line in a plane whose horizontal trace is $PQ$ (Prob. VII.) Its true length will then be shown on **H** by revolving it into **H**, about $PQ$ as an axis. After such revolution, $aa'$ will be found at $a''$, on a perpendicular $ra''$ to $PQ$ through $a$, and at a distance from $PQ$ equal to the hypothenuse of a right-angled triangle whose base is $ar$, and altitude $na'$. Finding $b''$ in a similar manner, $a''b''$ is the true length of $ab$—$a'b'$.

When the axis chosen is a projection of the given line, as at $a'b'$, Fig. 32, the plane last described becomes a projecting plane, the revolution is one of 90° and the hypothenuse coincides with the altitude described. Thus, on perpendiculars to $a'b'$ at $a'$ and $b'$, make $a'a_2 = an$, and $b'b_2 = bm$, giving $a_2b_2$ as the true length of $ab$—$a'b'$ after revolution about its vertical projection $a'b'$ to coincide with **V**. This solution (see Prob. VI.) illustrates (29, *Third*).

EXAMPLES.—1°. Revolve $ab$—$a'b'$, Fig. 32, about a perpendicular to **V** at $b'$ till parallel to **H**.
Ex. 2°. Revolve $ab$—$a'b'$, Fig. 32, about *any* vertical axis, not intersecting $ab$—$a'b'$, till the latter shall be parallel to **V**.
Ex. 3°. Revolve $ab$—$a'b'$, Fig. 32, about $ab$ as an axis till it coincides with **H**.
Ex. 4°. Revolve $ab$—$a'b'$, Fig. 33, about the vertical trace of any oblique plane containing it, till it coincides with **V**.
Ex. 5°. Find the true length of the perpendicular from a given point to a given plane.

## PROBLEM XIII.—*To find the shortest distance from a given point to a given line.*

*In Space.*—By Prob. IX., case 5°, pass a plane through the given point and perpendicular to the given line. Then, by Prob. XI., find where the given line pierces this plane. The line joining this point with the given point will be the line required, and its true length may be found by the last problem.

*In Projection.*—Pl. IV., Fig. 34. Let $pp'$ be the given point, and $ab$—$a'b'$ the given line. Knowing the directions of the traces of the perpendicular plane, this plane is determined by the line $pt$—$p't'$, where $p't'$ is perpendicular to $a'b'$, and $pt$ is parallel to $GL$. Then $PQ$ is drawn through the trace $t$, perpendicular to $ab$, and $QP'$ through $Q$, perpendicular to $a'b'$. The perpendicular plane $PQP'$ thus found intersects the given line at $qq'$, giving $pq$—$p'q'$ for the required distance. Its true length is $q'p_1'$. This solution accords with (29, *First*).

The usual solution, *in which the true length is found first* (28), is, to pass a plane through both the given point and the given line (Prob. IX., 2°), and then to revolve it about one of its traces into the plane of projection containing that trace, as in Pl. IV., Fig. 33, when the given point and line being seen in their real relative position, a perpendicular from the one to the other can immediately be drawn in its true length (29, *Second*).

EXAMPLES.—1°. Construct the solution just described.
Ex. 2°. Construct Fig. 34, with $pp'$ in any other angle than the *fourth*.
Ex. 3°. Construct Fig. 34, with $ab$—$a'b'$ crossing any other angle than the *second*.

**PROBLEM XIV.**—*To find the shortest distance between two lines not in the same plane.*

*In Space.*—The usual solution, illustrating (29, *First*) and readily constructed from a detailed description, is this.* Let $AB$ and $CD$ be the given lines. Then—

1°. By Prob. IX., 4°, pass a plane through the line $CD$ and parallel to the line $AB$. Call this plane $PQP'$.

2°. Project the line $AB$ upon the plane $PQP'$; that is, assume any point $M$ on $AB$, and by (Theor. II.) draw from it a perpendicular $XY$ to the plane $PQP'$; find, by Prob. XI., the point $N$, where $XY$ pierces $PQP'$; and through $N$ draw a parallel, $A_1B_1$, to $AB$, which will be the projection of $AB$ upon the plane $PQP'$.

3°. Note the point $I$ where the projection $A_1B_1$ of $AB$ upon $PQP'$ intersects $CD$.

4°. At $I$ draw a parallel to $XY$ and it will intersect $AB$ at a point $K$, and $IK$ will be the required shortest distance and common perpendicular between the two lines; for $I$ is where the projecting plane of $AB$ upon $PQP'$ cuts $CD$.

5°. By Prob. XII., find the true length of this common perpendicular.

In constructing this description, it only remains to

---

* In all constructions thus made from description, it is very important that the student should make the successive steps of the construction, one by one, as he reads their description; that is, he should not attempt to imagine the whole construction from the whole description, before beginning to draw it.

denote by corresponding small letters, $ab$—$a'b'$, $mm'$, etc., the *projections* of every point and line above designated in *space* by capitals, $AB$, $M$, etc.

*Otherwise*, and illustrating (29, *Second, Third*) a new plane of projection, a vertical one $V_1$, for example, may be taken through or parallel to either of the given lines, as $AB$. The system, composed of the two given lines, may then be revolved about any axis perpendicular to $V_1$, till $AB$ takes a vertical position, when the required shortest distance being horizontal will immediately appear by its projection on **H**.

*In Projection.*—Pl. IV., Fig. 35. Let $ab$—$a'b'$ and $cd$—$c'd'$ be the two given lines. Make $ab$ the ground line of a new vertical plane containing $ab$—$a'b'$, which line will therefore, by Prob. VI., appear projected upon this plane, when the latter is revolved into **H**, at $a''b''$. In like manner, $cd$—$c'd'$ will appear at $cd$—$c''d''$. Then revolve $ab$—$a''b''$ and $cd$—$c''d''$, together, about an axis perpendicular to the new vertical plane, as at $m''$, until $ab$—$a''b''$ becomes vertical at $b_1$—$b_1''m''$, when $cd$—$c''d''$, revolving to an equal angular extent—$c''m''c_1'' = b''m''b_1''$—about the same axis, will appear at $mc_1$—$m''c_1''$, since $mm''$, being in the axis, remains fixed. The line $ab$—$a'b'$ being now brought to the vertical position, $b_1$—$b_1''m''$, all perpendiculars to it will be horizontal, and by (Theor. I.) that one of them which is also perpendicular to $mc_1$—$m''c_1''$ will appear so in horizontal projection; hence $n_1p_1$, perpendicular to $mc_1$ from $b_1$, is the true length of the required perpendicular or shortest distance between the two lines. Its revolved new vertical projection is $n_1''p_1''$ (parallel to $ab$ as a ground line), which, by counter-revolution around the axis $mn$—$m''$, appears at $np$—$n''p''$, and thence on the primitive planes at $np$—$n'p'$.

We thus see that by the methods illustrated in Fig. 35, proceeding in an inverse order, we *first* find the desired true distance, and thence, by *counter-revolution*, and a *return to the primitive planes*, the projections of its original position, if these be desired; while by the method of the first solution, we proceed directly by first finding the projections of the primitive position of the required distance, and thence its true length.*

EXAMPLES.—1°. Let the given lines cross different angles.

Ex. 2°. Let both projections of each make angles of 30°, or less, with the ground line.

Ex. 3°. Let both projections of each make angles of 60°, or more, with the ground line.

Ex. 4°. Let one of the lines be the ground line.

Ex. 5°. Find the required shortest distance by Ex. 3, Prob. XI., 3°; the given line $A$ being in this example perpendicular to the directions of both of the given lines $C$ and $D$, and hence taken as the intersection of two planes each assumed by (Theor. II.) so as to be perpendicular to one of the given lines $C$ and $D$.

## PROBLEM XV.—*To find the true size of the angle between two given lines.*

*In Space.*—Find either trace of the plane containing the given lines (Prob. XI., 3°), then, by (Prob. XII.), revolve this plane into the plane of projection containing that trace, when the given lines will there show their real relative position.

*In Projection.*—Pl. IV., Fig. 36. Let $ab$—$a'b'$ and $ac$—$a'c'$ be the given lines, and hence $bac$—$b'a'c'$ the projections of the required angle. By Prob. VIII., find $p$ and $q$, the horizontal traces of the lines, and $pq$ will be the like trace of their plane. Then revolve this plane about this trace until it coincides with **H**, when $aa'$ will appear at $a''$ on $ar$, the horizontal trace of the plane of its revolution (27, 4°), and at a distance, $a''r$, from $pq$, equal to the hypothenuse of which $ar$ and $a't$ are the base and altitude. Since $p$ and $q$, being in the axis, remain fixed, $pa''q$ is then the true size, shown on **H**, of the given angle.

EXAMPLES.—1°. Let the projections of the given angle be obtuse.

Ex. 2°. Revolve the plane of the angle about its vertical trace and into **V**.

Ex. 3°. Let the vertex of the angle be in any other angle than the *first*.

---

* See my larger DESCRIPTIVE GEOMETRY (New York, 1874), where a number of problems in which the inverse order is followed are conveniently classed under the title, "*Inverse or counter-development.*"

Ex. 4°. Let one of the given lines be parallel to the ground line.
Ex. 5°. Let the two lines be the two traces of a plane.

**PROBLEM XVI.**—*To bisect a given angle in space.*

*In Space.*—Equal arcs subtend equal angles, but, if in space, may be differently inclined to a plane of projection, and hence unequal in projection; hence, if an angle be equally divided in space, its projections will not be so divided, unless the plane of the angle be parallel to a plane of projection, to which position it must therefore be brought by revolution about a suitable axis.

*In Projection.*—1°. *The general case.* Pl. IV., Fig. 36. Here both sides of the angle $bac$—$b'a'c'$ are oblique to both planes of projection. Then finding first the real size, $pa''q$, of the angle by the last problem, bisect it by the line $a''n$, and project $n$ at $n'$. In counter-revolution, $a''$ returns to $aa'$, and $nn'$, being in the axis, remains fixed, and $na$ and $n'a'$ become the *projections* of the bisecting line $na''$.

2°. *Special case.*—Pl. V., Fig. 37. In this case, which oftener occurs in practice, one side, $bc$—$b'$, of the angle, $abc$—$a'b'c'$, is perpendicular to a plane of projection, in this case to **V**. Revolve the angle about $bc$—$b'$ as an axis till it is parallel to **H**, as shown by revolving any point, as $aa'$, of the side $ab$—$a'b'$ in the arc $aa_1$—$a'a_1'$, where $b'$ is the centre of $a'a_1'$, till $ab$—$a'b'$ becomes horizontal as at $a_1b$—$a_1'b'$. Then bisect the revolved angle $a_1bc$ by $d_1b$, whose vertical projection is $d_1'b'$, and in counter-revolution, $a_1b$—$a_1'b'$, returns, as shown, to $ab$—$a'b'$, and the bisecting line similarly to $db$—$d'b'$, which gives the *projections* of the semi-angles.

EXAMPLES.—1°. Let one of the sides of the angle be vertical.
Ex. 2°. Let one of the sides be parallel to the ground line.

**PROBLEM XVII.**—*To find the true size of the angle included between two given planes.*

*In Space.*—Principles, leading to variously modified constructions, are as follows (Pl. V., Fig. 41): *First.* The plane, *mno*, of the required angle will be perpendicular to the intersection, *PP'*, of the two planes. *Second.* Hence by (Theor. II.) the traces, as *mn*, of this plane will be perpendicular to the like projections, as *Pb*, of the intersection; and the sides, *mo* and *no*, of the angle will be perpendicular to the intersection, *PP'*, of the planes. *Third.* As the plane *mno* is perpendicular to *PP'*, it will be perpendicular to every plane through that line, and hence to its projecting plane *P'bP;* but *P'bP* being perpendicular both to **H** and to *mno*, is perpendicular to their intersection *mn*. Hence *or*, the intersection of these two planes, is perpendicular to *PP'*, as being in the plane *mno*, and also to *mn* as being in the plane *P'bP*.

When one of the two given planes is a plane of projection, the plane of the required angle is perpendicular to the *trace* of the given plane on that plane of projection, and hence to the latter plane itself.

*In Projection.*—1°. *One of the given planes a plane of projection.* Pl. V., Figs. 38, 39. In Fig. 38, let the angle between **H** and the plane *PQP'* be required. Then assume any auxiliary vertical plane *PoP'*, whose horizontal trace, *Po*, is perpendicular to *PQ*. The lines *Po* and one in the given plane from *P* to *P'* are the sides of the required angle. Hence, by revolving this angle about *P'o* as an axis till it coincides with **V**, as indicated by the arc *Pn* whose centre is *o*, the angle is shown in its real size at *P'no*. Fig. 39 shows the case in which the given plane *PQP'* has the most general kind of position. Briefly, *Pr*, perpendicular to *GL*, is the *horizontal*, and *rs'*, perpendicular to *P'Q*, is the *vertical* trace of the plane perpendicular both to **V** and *PQP'*, and therefore containing the angle between *PQP'* and **V**. Revolving this plane about *Pr* and into **H**, *Ps''r* shows this angle in its true size.

*oP''p* shows the angle between *PQP'* and **H**, but does so by revolving the plane *P''oP'* of the angle about its horizontal trace into **H**, by making *op* = *oP'* and perpendicular to *oP''*; instead of by revolving it about its vertical trace *oP'*, as in Fig. 38.

2°. *The general case without auxiliary developments* (29, *First*).—Pl. V., Fig. 40. *PQP'* and *PRP'* being the given planes whose included angle is required, *Pb—P'a'* is their intersection. Then assuming any point *o*, on the ground line, by (*Prin. Second*) draw *on* perpendicular to *Pb*, and *oc'* perpendicular to *P'a'*, and *noc'* will be the plane of the required angle. By Prob. XI., this plane cuts the intersection *Pb—P'a'* of the two planes at *dd'*, which point, joined with *nn'* and *kk'*, where the plane *noc'* cuts the horizontal traces of the given planes, gives *ndk—n'd'k'*, the *projections* of the required angle, whose *true size*, shown at *nDk*, is found precisely as in Prob. XV., *Dr* being the hypothenuse formed on *dr* and *d'f'* as a base and altitude.

3°. *With auxiliary developments.*—Pl. V., Fig. 42. Let *PQP'* and *PRP'* be the given planes. *PbP'* is the projecting plane of their intersection upon **H**, and *Pb* is therefore the horizontal projection of that intersection. Revolving *P* about *b* to *P''*, and drawing *P''P'*, this line is the intersection *PP'*, in space, revolved about *P'b* to coincide with **V**. Then *bd''*, perpendicular to *P'P''* (see *ro*, Fig. 41), is the true perpendicular distance of the vertex *d''* of the required angle from the horizontal trace *nk* of its plane, *nk* being perpendicular to *Pb* at *b*. When the plane of the angle revolves about *nk* into **H**, *d''* will appear at *d*, by making *bd* on *bP* equal to *bd''*. Hence *ndk* is the required angle between *PQP'* and *PRP'*.

EXAMPLES.—1°. Let the traces of the given planes on one of the planes of projection be parallel.

Ex. 2°. Let one of the planes be perpendicular to the ground line.

Ex. 3°. Let the plane in Fig. 40, the construction remaining the same, be situated as in Fig. 42.

Ex. 4°. In Fig. 42, revolve the vertical projecting plane *PbP'* about its horizontal trace.

Ex. 5°. Let both planes be parallel to the ground line.

Ex. 6°. Let both planes be nearly perpendicular to the ground line.

Ex. 7°. Find the bisecting plane of the angle between the given planes. [In Fig. 42, for example, the horizontal trace of this plane would join *P* with the point where the bisecting line of the angle *ndk* meets *nk*. Its vertical trace would pass through *P'*].

## PROBLEM XVIII.—*To find the angle made by a line with a plane.*

*In Space.*—The angle made by a line with a plane is that made by the line with its projection on the plane. But this projection, the line itself, and the perpendicular which projects any one of its points upon the plane, form a right-angled triangle (Prob. XII.), in which the angle made by the given line with the projecting perpendicular is the complement of the required angle, which last can thence be found without finding the projection of the line upon the plane.

*Otherwise,* by rotating the system of the given line and plane, till, for example, the plane should be vertical, the perpendicular to it from a point of the given line would be horizontal, and hence a convenient axis about which to again revolve the given line till it should be horizontal, when the triangle above mentioned would be immediately known in its real size.

*In Projection.*—By the method of rotation (29; *Second*). Pl. V., Fig. 43. Let $PQP'$ be the given plane, and $ab$—$a'b'$ the given line. Assuming any convenient plane, $PRS'$, perpendicular both to $P'Q$ and **V**, revolve it about its trace $PR$ as an axis into **H**, as indicated by the arc $S'Q_1$, when $P'Q$, revolving with $PRS'$, will appear at $Q_1P_1$ perpendicular to $GL$. The point $P$, being in the axis $PR$, remains fixed, and $PQ_1P_1$ is the given plane after revolution to a vertical position. By revolving $ab$—$a'b'$ an equal angular extent about the same axis, the relative position of the line and plane will be preserved. This is indicated by the arcs $b'b_1'$ and $a'a_1'$, both having $R$ as their centre and being of equal angular extent with $S'Q_1$. The horizontal projections of these arcs are $bb_1$ and $aa_1$, parallel to $GL$, and giving $a_1b_1$—$a_1'b_1'$ for the revolved position of the given line. This done, the horizontal $a_1p$—$a_1'p'$, where $a_1p$ is perpendicular to $PQ_1$, is the perpendicular from $a_1a_1'$ to the revolved plane $PQ_1P_1$, meeting it at $p$, as $a_1b_1$ does at $c$. Hence, by making $pc''$ equal to the hypothenuse formed on $pc$ and $n'c'$

as a base and altitude, and drawing $c''a_1$, we have the required angle, $pc''a_1$.

*By the usual solution,* we should have found the horizontal traces of $ab$—$a'b'$ and of a perpendicular from $aa'$ to $PQP'$, and should then have proceeded, as in Prob. XV., to find the angle at $aa'$ between these two lines, which would have been, as above, the complement of the required angle.

EXAMPLES.—1°. Construct the solution just indicated.
Ex. 2°. Let the plane be situated as in Fig. 39.
Ex. 3°. Let the plane be parallel to the ground line.
Ex. 4°. Substitute for the given plane, the planes of projection.
[Find both traces of the line, then, recalling that the angle made by a line with a plane of projection is the angle made by it with its projection on that plane, by revolving the line *each way*, about *each trace*, of *each* of its two projecting planes, we can show its *true length* eight times, as well as each of the required angles four times. In Fig. 32, $b'a_1'a'$ is the angle between $ab$—$a'b'$ and H, and that between $a_2b_2$ and $a'b'$ the one between $ab$—$a'b'$ and V. See also $P'no$, Fig. 38; $oP''p$ and $Ps''r$, Fig. 39, and $P'P''b$, Fig. 42, for examples of the angles made by lines with the planes of projection.]

## PROBLEM XIX.—*To reduce an angle to the horizon.*

*In Space.*—The enunciation expresses the operation of finding the horizontal projection of an angle in space, given by the inclinations of its sides to each other and to a vertical line. These inclinations may be shown by auxiliary developments.

*In Projection.*—Pl. V., Fig. 44. Let $ao'o$ and $b_1o'o$ be the angles made by the two sides of the given angle with a vertical line $oo'$, and let $b_1o'b_2$ be the true size of the angle shown by revolving it about $o'b_1$ into V. Then $oa$ being the horizontal projection of $o'a$, the horizontal projection, $ob$, of $o'b_1$ will be found by the arcs $b_1b$ with $o$ as a centre, and $b_2b$ with $a$ as a centre, which intersect at $b$, and $aob$ is the horizontal projection of the angle $b_1o'b_2$.

## PROBLEM XX.—*To construct the projections of a plane figure lying in a given plane.*

*In Space.*—Determine a sufficient number of points of the figure, given by its development, by parallels or perpendiculars to the traces of its plane, and transfer these points by counter-revolution to the projections of the same lines.

*In Projection.*—1°. *Preliminaries.* Pl. V., Fig. 45. Let the given figure be a circle, and let $PQP'$ be its plane. Assuming any plane, $PRR'$, perpendicular to the vertical trace $P'Q$, revolve the point $P$ of the horizontal trace, in this plane, to $P_1$, by making $P_1$ the intersection of an arc of radius $QP_1$, equal to $QP$, with $RR'P_1$, the vertical trace of the plane of the revolution. Then $P'QP_1$ will be the plane $PQP'$ revolved about $P'Q$ into **V**, and the circle with $o''$, assumed as a centre, will be the revolved position of the given circle.

2°. *Projections of points of the circle.*—Draw $o''s'$, for example, parallel to $P_1Q$. Its vertical trace is $s'$ on $P'Q$, and it is in space parallel to $PQ$; hence, projecting $s'$ at $s$, its projections are $s'o'$, parallel to $GL$, and $so$, parallel to $PQ$. By counter-revolution about $P'Q$, $o''$ appears at $o'$, the intersection of $o''o'$ perpendicular to $P'Q$, with $s'o'$, and thence, by projection, at $o$ on $so$, or $so = s'o''$. In like manner, any other points may be found.

*Again:* draw, for example, $o''r''$, parallel to $P'Q$. Making $Qr = Qr''$, we find $ro$, parallel to $GL$, for its horizontal projection; and, projecting $r$ at $r'$, its vertical projection is $r'o'$, parallel to $P'Q$. Then $o'o$ is found as before from $o''$, but on $r'o'$ and $ro$. Also $r'o' = r''o''$. Other points can be likewise found.

The points $aa'a''$ and $bb'b''$ are the highest and lowest, and are on $a''g''$ perpendicular to $P_1Q$, and whose horizontal projection, $ag$, is perpendicular to $PQ$, at $g$, found by making $Qg = Qg''$.

Producing $oo'$ to $k''$, and making $Qt'' = Qt$, the line $k''t''$ shows the revolved position of a line in $PQP'$, and whose projections are perpendicular to $GL$. Then drawing the tangents, as $f''d''$, parallel to $k''t''$, we find the tangents, as $dd'$, which include between them the two projections of the circle, and, as in the previous constructions, their contacts, as $d$ and $d'$, with these projections.

EXAMPLES. 1°.—Let the given plane be revolved into H.

Ex. 2°. Let it be in the kind of position shown at *PQP'*, Fig. 39.

Ex. 3°. Substitute for the circle a star or other plane figure.

Ex. 4°. Substitute for any plane figure a cube one face of which shall be in the given plane.

THEOREM III.—*The projections of a circle seen obliquely are ellipses.* Pl. V., Fig. 45.

*Parallels in space are parallel in projection, and lines equally divided in space are equally divided in projection.* Thus every line, as *a'b'*, through the centre of either projection, has a pair of parallel tangents at its extremities, as at *a'* and *b'*, and bisects all the chords, as *e'c'*, which are parallel to these tangents. One of these chords, *c'e'*, passes through the centre *o'*, and hence is a *diameter*. Two diameters, as *a'b'* and *c'c'*, each of which, as in the original circle, bisects chords parallel to the other, are called *conjugate diameters*. Also, in each projection, two diameters, called the *axes*, are *perpendicular* to the parallel chords and tangents which they intersect, as seen at *ab* and *cc*. These properties belong to the ellipse; hence we conclude that the projections of a circle seen obliquely are ellipses.

The longer and shorter axes are distinguished as respectively the *transverse* and *conjugate*, or the *major* and *minor* axes.

## CHAPTER II.

### DEVELOPABLE SURFACES.

#### *Definitions.*

30. Proceeding with *ruled surfaces* (22), we come to *single curved surfaces*, of which *developable* surfaces, as the simplest, are considered first.

The only developable surface of revolution is the *cone*, including the cylinder of revolution as one of its extreme cases. *The cone of revolution* is generated by the revolution of a straight line about an axis which it intersects at any angle.

31. The opposite extreme cases of the cone are, *first*, that in which this angle is nothing, when the generatrix (19) is *parallel* to the axis, and the surface becomes a *cylinder* of revolution, and, *second*, that in which the same angle is a right angle, when the cone becomes a plane (23).

32. The intersection of the generatrix of a cone with its axis is called its *vertex*. As the generatrix is of indefinite extent in both directions from this point, in an immaterial or purely geometrical cone, the complete conic surface consists of two equal and opposite parts, of indefinite extent, on each side of the vertex. These are called the *two nappes*.

33. *The base* of a cone of revolution is any plane section of it, usually understood as a circular one, at which its ele-

ments are supposed to terminate, or at which it rests upon some plane.

Thus the horizontal trace of any cylinder or cone is also often called its base. Circular sections of surfaces of revolution generally, are called *parallels*, and are in planes perpendicular to the axis of revolution. Planes containing the axis are called *meridian planes*, and the sections in them, *meridians*.

### A—Projections.

34. As already seen (Theor. III.), the projections of circles seen obliquely are ellipses. Such, therefore, will be one or both of the projections of the base of a cylinder or cone whose axis is oblique to either **H** or **V**, or to both.

It may then evidently be convenient to be able to construct an ellipse when its *axes*, as *ab* and *ce*, Pl. V., Fig. 45, are known, or when a pair of *conjugate diameters* are known. It will also be convenient to construct a tangent to an ellipse, *first*, at *a given point*, since it is evidently easier to sketch a curve through given points when the tangents at those points are known; *second*, from a given exterior point, as from the projections of the vertex of a cone to those of its bases (see Pl. VI., Fig. 52); *third, parallel to a given line*, as at the limiting projecting lines, *dd'*, Pl. V., Fig. 45.

We will therefore begin with the two following preliminary general problems upon the ellipse, embracing these particulars.

PROBLEM XXI.—*To construct an ellipse under various given conditions.* Pl. VI., Figs. 46–48.

1°. *Construction of the ellipse on given axes by radials from the extremities of an axis.*—Fig. 46. Let $2AB$ and $CD$ be the given axes. Divide $AB$ and $Be$, which is equal and parallel to $AC$, into the same number of equal parts, similarly numbered, as shown, in reckoning to or from $B$.

Then radials from $C$ and $D$, through points similarly numbered on $Be$ and $AB$, respectively, will intersect at points, as 1, 2, 3, of the curve.

In like manner, each of the other three quadrants of the curve can be found.

*Theory.*—If $CA = AB$, the curve would become a circle, and by the same construction the triangles, $C_1D$, $C_2D$, etc., would then be right-angled at 1, 2, etc., on the curve, as they should be. Fig. 46 is then simply the projection of this construction of the circle, as seen after revolution about its diameter, equal and parallel in space to $AB$, until the diameter perpendicular to this one is shown as at $CD$.

2°. *Construction of the ellipse by concentric circles on the given axes.*—Fig. 47. Let $OA$ and $OC$ be the given semi-axes, on which describe circles, as shown, and divide the latter into corresponding parts, as $Ac$ and $Da$, by radii which may be equidistant if preferred. Then, from points of division, as $b$ and $d$, on the same radius, draw, from the former, parallels to $OA$, and from the latter, parallels to $OC$, whose intersections, as at $f$, will be points of the required ellipse.

*Theory.*—From the triangle, $Obg$, since $fd$ is parallel to $Og$, we have,

$$fg : bg :: Od : Ob$$
$$:: OA : OC.$$

The lines $fg$ and $bg$ being called ordinates of the ellipse and circle of radius $OC$, respectively, the first and last couplets of the above proportion express the property of the ellipse that the *ordinate to the ellipse* is to the corresponding *ordinate to the circle on its transverse axis*, as the *semi-conjugate axis* is to the *semi-transverse axis*.

The curve given by the construction is thus identified as an ellipse.

3°. *Construction of an ellipse on a given pair of conjugate diameters—First Method.*—Fig. 48. Let $ab$ and $cd$ (Theor. III.) be a pair of such diameters. Draw $fg$ through $c$ and parallel to $ab$. At $c$ erect a perpendicular, $cD$, to $fg$, make it equal to $ab$, and on it as a diameter describe the circle

*ADBc*. If now a circle be described passing through *O* and *o*, and having its diameter on *fg*, right angles, *fOg* and *fog*, can be inscribed in it, and the sides of the latter will be the required axes. The centre, *h*, of this circle is found by drawing the line *Oo*, and *ch* perpendicular to it at its middle point. The limits, *m* and *n*, of the semi-axes are found by drawing *Mm* and *Nn* parallel to *Oo*. Making $op = om$, and $oq = on$, gives the axes, having which the curve can be found as already described.

*Theory.*—Consider the paper above *fg* as a vertical plane in which lies the circle of radius *OA*; and the paper below *fg* as a horizontal plane on which this circle is projected by projecting lines, of which *Oo* is one. Then *cd* will be the projection of *cD*; *ab* equal and parallel to *AB*; and *mo* and *no*, being the projections of *MO* and *NO*, are not only perpendicular to each other at the centre, *o*, of the ellipse, but, like *MO* and *NO*, each bisects the chords which are parallel to the other (Theor. III.). Hence they are the axes.

*Second Method.*—Fig. 48. At *a* draw a parallel, *ax*, not shown, to *oc*, meeting *fg* at *x*, and completing the parallelogram *aocx*. Then a construction like that of Fig. 46, by lines from *c* and *d* to like points of division on *ax* and *ao* respectively, will give points of the arc, *amc*, of the ellipse. Having found the entire curve in this manner, the axes can be very simply found as follows: On any diameter, as *ab*, describe a semicircle which will cut the corresponding half of the ellipse in a point. Lines from this point to *a* and *b* will be chords including a right angle, they being thus inscribed in a semicircle; then lines through the centre, *o*, and parallel to these chords will be the required axes, by a property of the ellipse that supplementary chords—that is, chords from any point of the curve to the extremities of a diameter—are parallel to a pair of conjugate diameters, and, if perpendicular to each other, are therefore parallel to the *axes*.

*Theory.*—The figure thus formed is merely the projection of one like Fig. 46 upon a plane parallel to it, by means of projecting lines which are *oblique* to the plane of projection, instead of being perpendicular to it, as is usually the case.

EXAMPLES.—1°. In Fig. 46, find the other three quadrants of the ellipse.
Ex. 2°. In Fig. 46, when conjugate diameters, instead of axes, are given.
Ex. 3°. Having given the curve and a diameter, find the axes.

PROBLEM XXII.—*To construct tangents to a given ellipse.*

1°. *At a given point on the curve—First Method.*—Fig. 49. Let *ACB* be a semi-ellipse to which a tangent is to be drawn at *S*. The semicircle *AcB* may be considered as the position, after revolution about *AB* and into the plane of the paper, of the circle in space whose projection is *ACB*. Then *S* will be found at *s*, and *sT* will be the revolved position of the required tangent. The point *T* being fixed because in the axis, *s* returns by counter-revolution to *S*, and *TS* is the required tangent.

*Second Method.*—Let *R* be the given point. Revolve it to *r*, and draw a radius, *Os*, perpendicular to *Or*. Then, by Prob. XX., *OR* and *OS* will be a pair of semi-conjugate diameters. Hence, *Rc*, parallel to *OS*, is the required tangent at *R*.

2°. *Through a given exterior point.*—Fig. 50. As in Fig. 49, let *AcB* represent the revolved position of the circle whose horizontal projection is *ACB*, and let *P* be the given point in space, in the plane of the circle. Then projecting *c* at *c'* upon the ground line *GL*, the arc *c'C'*, with *A'* as a centre, is the vertical projection of *cC*, giving *A'C* as the vertical projection of the circle in space, and *P'*, on *A'C* produced, as the vertical projection of *P*, since, as *AB* is perpendicular to *GL*, the plane of the circle is perpendicular to **V**, and therefore *A'C* is its vertical trace. Revolving into **H** about *AB*—*A'* again, *PP'* appears at *pp'*, and *pt*, tangent from *p* to the semicircle at *t*, is the revolved position of the required tangent. In the final counter-revolution, *pp'* returns to *PP'*, and *t* to *T*, and *PT* is the required tangent.

A second tangent from *P* can be similarly found.

3°. *Parallel to a given line—First Method.*—Fig. 51. Proceeding by the same method as in Fig. 50, let $ACB$ be the given ellipse, and $nP$ the given line; regarded as the horizontal projection of a circle in space and a line in its plane. Then, revolving about the axis $AB$—$A'$, we find $A'C'$ and $P'$ as before, and, by revolving into **H** again, $P$ will appear at $p$, and $n$, being in the axis, remains fixed, giving $np$ for the revolved position of $nP$ into **H**, and hence $th$, parallel to $np$, for the revolved position of the given tangent. Counter-revolving again, $t$ returns to $T$, and $pp'$ to $PP'$, whence $TH$, parallel to $nP$, is the required tangent.

*Second Method.*—Draw a line, $OS$, not shown, parallel to $nP$, meeting the ellipse at $S$. By revolution, show its revolved position, $Os$, as in Fig. 49, and draw $Ot$ perpendicular to $Os$. Then, finding $T$ by counter-revolution, $TN$, parallel to $OS$ or to $nP$, will be the tangent required.

EXAMPLES.—1°. In Fig. 49, take the ellipse in other positions, and apply each method separately to different quadrants of it.

Ex. 2°. In Fig. 50, show the entire ellipse in a different position, change the position of $P$, and draw both of the tangents from it. [$GL$ must be perpendicular to $AB$.]

Ex. 3°. In Fig. 51, show the entire ellipse and vary the direction of $nP$.

35. The perfection of the form of a curve constructed by points depends less on the number of points than on the accuracy with which they are located. In the case of an ellipse, *eight points*, exactly located, *four being the extremities of the axes*, and *one other in each quadrant*, are almost always enough to enable one to sketch the curve through them.

PROBLEM XXIII.—*To represent a cone of given axis and diameter by the projections of its vertex, circular sections, and visible limits, the axis being oblique to both* **H** *and* **V**; *also any element or point of the surface.*

*In Space.*—A cone being determined by its vertex and any plane section not containing its vertex, its projections will consist of the projections of these parts, to which are usually added the projections on each plane of those elements which form the boundaries of the convex surface, as seen in looking perpendicularly at that plane.

These elements are called the *extreme elements*, and as all the others, and hence the projections of the given section on each plane, are included between these, their projections are tangent to those of the section.

*In Projection.*—Pl. VI., Fig. 52. 1°. *Four points of each projection of the section.* Let $Vo$—$V'o'$ be the axis of the cone, and $ab$ its diameter at the point $o$. This diameter is in space horizontal, and hence (Theor. I.) perpendicular to $Vo$, and, in vertical projection, parallel to $GL$, as at $a'b'$. That diameter at $oo'$ which is parallel to **V** is shown at $c'd'$, equal to $ab$, perpendicular to $V'o'$, and in horizontal projection parallel to $GL$, as at $ab$. We thus get four points, $aa'$, $bb'$, $cc'$, and $dd'$ of the section.

2°. *The shorter axis of each projection of the section.*—Assume a new vertical plane of projection, parallel to the axis of the cone, or, as at $G_1L_1$, containing it. By making $VV'' = V'y$, and perpendicular to $G_1L_1$, and $oo''$ likewise equal to the height of $o'$ above $GL$, the new projection of the axis will be $V''o''$; and $n''r'' = ab$, and perpendicular to $V''o''$, through $o''$, will be the new projection of that diameter of the cone which is in this vertical plane. Its horizontal projection is $nr$, and its vertical one, not drawn, is $n'r'$, where the heights of $n'$ and $r'$ above $GL$ are equal to those of $n''$ and $r''$ above $G_1L_1$ (Prob. VI.)

As these are evidently the highest and lowest points of the section, *their tangents* are horizontal, and hence, in horizontal projection, parallel to $ab$, and, in vertical projection, to $GL$.

*Again:* the shorter axis of the vertical projection being evidently in the plane perpendicular to **V** through $V'o'$, since it is perpendicular to $c'd'$, assume $G''L''$, parallel to $V'o'$ as the ground line, upon a vertical plane through $cd$—$c'd'$, for example, of a new plane of projection perpendicular to **V**. Then $o'''$ will be the projection of $oo'$ upon this plane, and making $a''p'' = ap$, we get $o'''a''$ as the projection of the given circular section upon the new plane, and, making $o'''e'' = oq$, we find the radius whose vertical projection is $o'e'$, and horizontal projection $oe$, where $et = e''t''$. The opposite extremity of this diameter is $ff'$.

3°. *Other points.*—We have now found *eight* points of the projections of the circular section, which, as four of them determine the axes in each projection, may serve in sketching the curve. But as they are in this case quite ununiformly distributed, four more may usefully be found in each projection by simply drawing in the horizontal projection the diameters symmetrical with $cd$ and $ef$, and therefore equal to them, and likewise, in the vertical projection, those symmetrical with $a'b'$ and $r'n'$, giving, in all, twelve points in each projection.

Finally, tangents to $arbf$ from $V$, and to $a'c'b'd'$ from $V'$, drawn by the eye, or found as in Fig. 50, for each projection separately, will complete the projections of the cone. In the latter case, two more points of each projection, making *fourteen* in all, will be known.

4°. *Partial plane construction of the circular section.*—By first finding the axes $ab$ and $rn$ of the horizontal projection, and $c'd'$ and $e'f'$ of the vertical projection, the curve in each case may be completed as a plane problem by either of the constructions shown in Prob. XXI., 1°, 2°.

Also, as $ab$ and $rn$ are perpendicular to each other in space, each bisects chords parallel to the other, and will do the like in all projections; hence, $a'b'$ and $r'n'$ are conjugate diameters, and in like manner so are $cd$ and $ef$. As soon, therefore, as only these diameters have been found, the respective projections of the curve may be completed by either method of Prob. XXI., 3°.

5°. *Visibility of the curve—Elements and points.*—In horizontal projection, that portion of the curve is visible which is between the extreme elements and on the upper half of the cone, as determined from the vertical projection by inspection. Thus, $dd'$ being on the upper half, the full portion of the curve is visible, and the dotted portion invisible, in horizontal projection.

. In vertical projection, the entire circumference of the circular section would be visible if the surface of the cone were limited by it, since the **vertex** is further from the observer looking towards **V** than the base is. But the conic surface being produced as shown, only the *front* part of the circular section is visible—that is, the part through *ee'*, and limited by the extreme elements from *V"*.

Finally, *Ve—V"e'* being any element, and being careful that *e* and *e'* are the projections of the same point, *kk'* will be a point on that element.

EXAMPLES.—1°. Construct a larger cone and with a larger angle at its vertex, so as to clearly distinguish the extreme elements, tangent to the circular section, from those through *aa'*, *dd'*, etc.

Ex. 2°. Let the axis of the cone be parallel to **H** or **V**.

Ex. 3°. Let it be nearly perpendicular to **H** or **V**.

### *The Conic Sections in General.*

36. A cone being often most conveniently represented by one or both of its *traces* (Prob. IV.), together with its extreme elements, the principal forms of its intersections with **H** and **V**, according to its position relative to them, are to be noted.

A plane can intersect a cone in only three essentially different ways. Thus, *VAB*, Pl. VI., Fig. 54, representing a cone whose axis, *VS*, and opposite elements, *VA* and *VB*, are in the plane of the paper, the intersecting plane may be supposed to be perpendicular to the paper. Then this plane may intersect the cone as follows:

1°. *Parallel*, as at *PQ*, *to a plane*, as *MN*, *which contains only the vertex* of the cone. In this case, the plane cuts all the elements of the cone, and the section is an *ellipse*.

2°. *Parallel to a tangent plane*, as the one of which *VA* is the trace. In this case the plane is parallel to the element *VA* only, and hence cuts all the elements but that one, and the section is called a *parabola*.

3°. *Parallel to a plane*, as the one whose trace is *Vm*, *which contains two elements* of the cone. In this case, the plane cuts both nappes (32) of the cone, and all the elements except the two in the plane $Vm$, and the curve, consisting of two opposite branches, is called a *hyperbola*.

37. *The ellipse*, Fig. 52, is a *closed curve*, since its plane cuts all the elements of the cone.

*The parabola and hyperbola* are *open curves*, and of infinite extent, since their planes, being parallel to certain elements, intersect them only at an infinite distance from the vertex, where, since the elements diverge from the vertex, they are also at an infinite distance apart.

38. *Special or elementary forms.*—When the plane $PQ$, Fig. 54, coincides with $MN$, the ellipse reduces to a *point*. When it is perpendicular to the axis $VS$ of the cone, the ellipse is a *circle*.

When the plane of the parabola coincides with the tangent plane at $VA$, the parabola becomes a straight line. When the cone, by the infinite distance of its vertex, becomes a cylinder, any intersecting plane parallel to a tangent plane cuts it in two *parallel straight lines*, which is thus another special case of the parabola.

Finally, when the plane of the hyperbola coincides with that whose trace is $Vm$, the hyperbola becomes a *pair of intersecting straight lines*.

Thus the *point, straight line, parallel lines, intersecting lines*, and *circle*, which, with figures composed of them, form the subject of elementary geometry, are simply the special and simplest forms of the conic sections.

39. *Curves determined by their projections.*—The projecting lines from all the points of a curve to a plane of projection, form a cylindrical surface of which that curve is the *directrix* (19), unless the curve lies wholly in a plane which is perpendicular to that plane of projection. The trace of this cylinder upon that plane will be the projection of the given curve, and, indeed, of *any* curve traced upon the convex surface of this cylinder. But, as in the case of a point

or line, the projecting cylinders erected on *two* given projections of a curve will intersect each other only in that *one* curve, which is thus determined by those projections.

THEOREM IV.—*In the ellipse, section of a cone made by a plane actually cutting all its elements, the sum of the distances from any point of the curve to two fixed points within it is constant, and equal to the transverse axis.*

Let $VAB$, Pl. VI., Fig. 54, represent a cone whose axis, $VS$, and opposite elements, $VA$ and $VB$, are in the plane of the paper; and let $PQ$ be the trace of a plane perpendicular to the paper and oblique to $VS$ in the manner stated. Then the circles of centres $S$ and $s$, in the cone's axis, and each tangent to $PQ$, $VB$, and $VA$, are the projections of two inscribed tangent spheres having circles of contact with the cone of which $AB$ and $ab$ respectively are chords. These spheres are tangent to the plane $PQ$ at $F$ and $F_1$. All these points and lines, with $SA$, $SF$, and $sF_1$ being in the plane of the paper, conceive the projecting lines employed to be oblique to the paper, and the curve $rpq$ may represent the section of the surface of the cone made by the plane $PQ$, and $ACB$ and $acb$ the circles of contact of the spheres with the cone.

This done, assume any point $p$ of the curve, and through it draw the element $VC$ of the cone, $pF$, and $pd$, which represents a perpendicular from $p$ to $PQ$. Then, since all tangents to a sphere from the same point are equal, and portions of elements of the cone between parallel circles are also equal,

$$Ar + ra = Fr + rF_1 = Cp + pc = pF + pF_1$$

also

$$bq + qB = F_1q + qF = Cp + pc = pF + pF_1$$

The third and fourth members of these equations being the same, their second members are equal; then, subtracting from the latter the common part $FF_1$, we have

$$2\,Fr = 2\,F_1q$$

whence, adding $FF_1$ again,
$$rF_1 = qF$$
hence from the first two equations,
$$pF + pF_1 = Fr + Fq = rq.$$

That is, the curve $rpq$ is such that the sum of the distances from any one of its points to two fixed points, $F$ and $F_1$, within it, is constant, and equal to its longest chord, $rq$. This property forms the usual definition of the ellipse; $rq$ is the transverse axis, while the fixed points $F$ and $F_1$ are called its *foci*.

As the demonstration takes no account of the angle $AVB$ at the vertex, we conclude that it is independent of it, and applies to the cylinder as a particular case of the cone, and hence that *the oblique section of a cylinder of revolution is an ellipse.*

40. *To construct an ellipse* by (Theor. IV.).—Lay down $rq$ and the points $F$ and $F_1$ anywhere between its extremities, but equidistant from them. With $F$ and $F_1$ successively as centres, describe arcs, first with any radius greater than $Fr$, and then with the remainder of $rq$ as a radius. The intersections of the latter arcs with the former will give four points of the curve.

These points, thus found in sets of four, will be seen to be symmetrically placed relative to $rq$ and a perpendicular to it at its middle point. Hence the ellipse, section of the cone, has two axes of symmetry, and is identical with the ellipse, as defined in Prob. XX.

41. By methods * quite like those of *Theorem* IV., it may be shown that a *parabola* is a curve such that the distances of any one of its points from a fixed point within, the *focus*, and from a fixed line, the *directrix*, are equal; that the *hyperbola* is distinguished by the property that the *difference* of the distances from any one of its points to two fixed points, *foci*, is constant and equal to the shortest chord connecting

---

* See OLIVIER, Cours de Géométrie Descriptive.

the two branches of the curve; also that the ellipse and hyperbola have *directrices*.

PROBLEM XXIV.—*To represent a cone by the projections of its vertex, horizontal trace, and extreme elements.*

*In Space.*—We may proceed in two ways: *first*, by finding any convenient number of the fourteen points of some circular section, by the last problem, and then finding the horizontal traces of the elements through these points; or, *second*, by first finding the axes of the horizontal trace, which will usually be an *ellipse* (36), and then completing the curve by Prob. XXI. (1°, 2°).

By an extension of (27, 4°) either trace of any element of a cylinder, or a cone, is in the like trace of that surface.

*In Projection.*—1°. *Without the conjugate axis.* Having a sufficient number of the points, $aa'$, $bb'$, $cc'$, etc., already described (Pl. VI., Fig. 52), according to the *scale* of the figure or the *distribution* of the points, depending on the position of the cone, then, without sketching the projections of the circular section through them, find the horizontal traces, as $A$, of $Vd$—$V'd'$ (taken to represent the extreme element which is so nearly confounded with $Vd'$), $D$, of $Vb$—$V'b'$, etc., and join these points, which will give the horizontal trace, $ADB$, of the cone, tangents to which, from $V$ (Prob. XXII., Fig. 50), will complete the horizontal projection of the cone.

Tangents as $AA'$ to the horizontal trace, and perpendicular to $GL$, found by Prob. XXII., Fig. 51, will project this trace on the ground line at $A'B'$, whence $A'V'$ and $B'V'$ will complete the vertical projection of the cone.

Since $E'$, vertical projection of $u$, does not bisect $A'B'$, we see that the *centre of the trace* is not in the axis $Vu$—$V'o'E'$ of the cone.

2°. *By the immediate construction of the axes of the trace.*—The *transverse axis*, $MR$, of the trace may be found either by finding the horizontal traces $M$ and $R$ of the elements $Vn$—$V'n'$ and $Vr$—$V'r'$, or by means of the auxiliary projection

$V''r''n''$, in which $V''r''$ and $V''n''$ intersect $G_1L_1$ in the same traces.

The conjugate axis, $PQ$, of the trace is thus found. Produce the axis $V''o''$ of the cone, and draw a perpendicular, $NC$, to it through $O$, the middle point of $MR$, and hence the centre of the trace of the cone; then $CN$ is the radius of that circular section of the cone whose plane intersects **H** in the line $PQ$, perpendicular to $MH$. The required conjugate axis of the trace is thus that chord of this circle which is perpendicular to the auxiliary vertical plane at $O$. The arc $NS$ of radius $CN$ is the revolved position of an arc of this circle, and $OS$, perpendicular to $CN$, is half of the desired chord; then making $OP$ and $OQ$ each equal to $OS$, we have $MR$ and $PQ$ the axes of the cone's trace, from which it can be constructed by Prob. XXI., 1°, 2°; or by (40), since, as the lines from $P$ or $Q$ to the foci are equal to each other, and their sum equal to $MR$, each is equal to $MO$. Hence find the *foci* by intersecting $MR$ by an arc from $P$ or $Q$ as a centre, and $MO$ as a radius, and then proceed by (40).

EXAMPLES.—1°. Let $VV'$ be in any other angle than the *first*.

Ex. 2°. Let the axis of the cone cross any other angle than the *first*.

Ex. 3°. Substitute a cylinder for a cone in each of the foregoing examples.

Ex. 4°. The axis of the cone being parallel to **V**, let one of the elements be parallel to **H**. [The horizontal trace will in this case be a parabola.]

Ex. 5°. The axis being as in Ex. 4, let a horizontal plane through the vertex contain two elements. [The horizontal trace will then be a hyperbola.]

**PROBLEM XXV.**—*To construct the projections of a cylinder of revolution having a given horizontal trace, or base; also any element and point of the surface.*

*In Space.*—From the symmetry of the cylinder, and its base, relative to a vertical plane through its axis, its extreme elements, in horizontal projection, will be tangents to the base at the extremities of its shorter axis, which is the diameter of the cylinder. Those seen in vertical projection will be parallels from the extremities of the vertical projection of the base, which will be on the ground line; and they will be at a distance apart equal to the diameter of the cylinder. All the elements are parallel in each projection.

*In Projection.*—Pl. VI., Fig. 53. Let the ellipse $ADBC$ be the given trace. Then, as described, $CI$ and $DK$, tangent at the extremities of the shorter axis $CD$, complete the horizontal projection of the cylinder. $CBD$ is dotted, being invisible.

Drawing the tangent projecting lines as $EE'$ (Prob. XXII., 3°), $E'F'$ is the vertical projection of the base, and $O'$ that of its centre $O$. Next, describe the circle $k't'$ of centre $O'$ and radius $OC$ that of the cylinder, and the extreme elements in vertical projection will be tangents to this circle from $E'$ and $F'$. To draw these important lines, to which all the other vertical projections of elements are parallel, some other points are provided. Thus, having, by the usual method, shown in the figure, found $k'$, the point of contact of a tangent from $E'$ to the circle centred at $O'$, make $MN$, parallel to $O'k'$, a fourth proportional, $MN$, to $E'O'$, $O'k'$, and any assumed distance $E'M$. Also making $E'O'' = E'O'$ and $O''k''$ equal and parallel to $O'k'$, we have four points, $k''$, $E'$, $k'$ and $N$, which, if all in the same straight line, will determine it very exactly. $F't'$, parallel to $k''E'k'$, is then the opposite element.

Finally, $Hp$, parallel to $CI$, being the horizontal projection of any element, and $p$, of a point upon it, project $H$ at $H'$, draw $H'p'$ parallel to $E'N$, and project $p$ upon it at $p'$, and $Hp$—$H'p'$, and $pp'$, will be the projections of an element and of a point upon it.

EXAMPLES.—1°. Vary the problem, beginning by drawing $CI$ and $DK$ in the opposite directions to the present ones from $C$ and $D$.

Ex. 2°. Make the application of Prob. XXI., 3°, in drawing the tangents $EE'$ and $FF'$.

PROBLEM XXVI.—*To construct the projections of a cone of revolution having a given elliptical trace or base.*

*In Space.*—Observing from Theorem IV., Fig. 54, that a sphere inscribed in a cone of revolution and tangent to the plane of an elliptical section, is so tangent at a focus of that section; and that by producing or reducing $FS$ or $F_{,}s$, other cones of revolution can be found having the same elliptical section $rpq$, we construct the two projections of any sphere tangent at either focus of the given trace. The required cone will then be tangent to this sphere.

*In Projection.*—Pl. VI., Fig. 55. Let $ACBD$ be the given elliptical horizontal trace of a cone of revolution, with the axis $AB$ parallel to the ground line. At either focus, as $S$, $F'$, of the trace, erect a perpendicular, and on it take any convenient radius, $S'F'$, and with it describe the equal circles at $S$ and $S'$, which are the projections of the tangent sphere at $S$, $F'$. Then, tangents from $A'$ and $B'$ to the circle $S'$ will complete the vertical projection of the cone, and, projecting $V'$ at $V$, on $AB$ produced, tangents from $V$ to circle $S$, which will also be tangent to the ellipse $ACBD$, will complete its horizontal projection.

The sphere of radius $S'F'$ may be tangent at either focus, and either above or below **H**, giving four cones.

EXAMPLES.—1°. Let $AB$ be oblique to $GL$.

Ex. 2°. Construct the *cylinder* of revolution whose horizontal trace is $ACBD$. [That there will be one such cylinder among the series of cones of which that in the figure is one, appears from the fact that when the variable circle $S'$ is larger than a certain size, the tangents to it from $A'$ and $B'$ will converge *downward*. There must, therefore, be one size of this circle which will make these tangents *parallel*. It may easily be shown by elementary geometry that the centre of this circle will be at the intersection of $FS'$ produced with the semicircle on $A'B'$ as a diameter.]

PROBLEM XXVII.—*To construct the projections of a cone whose axis is parallel to the ground line, and of any of its elements, and points.*

*In Space.*—These projections will evidently be two equal isosceles triangles. The construction of any element requires that of any point on the circumference of the base, whose plane is perpendicular to $GL$. This is accomplished by revolving the base about any suitable axis until it becomes parallel to or coincides with a plane of projection.

*In Projection.*—Let $VAD$—$V''B'C'$, Pl. VI., Fig. 56, be the projections of the cone. $AD$—$A'O'$ is the *horizontal*, and $OB$—$C'B'$ the *vertical* diameter of its base. By revolving this base about the former diameter, or any parallel to it, as an axis, the base will become parallel to **H**, and will then show its circular form in the horizontal projection. By revolving it about $OB$—$C'B'$, or any parallel to this line, it will likewise show its circular form in vertical projection. Then let $mV$ be the horizontal projection of any element; revolving the base 90° to the left about $AD$—$A'$ as an axis, its upper half will appear as at $AB''D$—$A'B'''$, and $m$ describes the quadrant whose horizontal projection is $mm''$ and will appear at $m''m'''$. By counter-revolution, this point returns in the arc $m''m$—$m'''m'$, and $m'V''$ is the vertical projection of the element $mV$; but as the cone is symmetrical relative to every plane through the axis, make $A'r' = A'm'$, and $r'V''$ will be another element whose horizontal projection is $mV$. The point $pp'$ is on the element $mV$—$m'V''$.

Again: let $a'V''$ be the given projection of an element. Revolving as before, $a'$ will appear at $a'''$, whence it is projected at $a''$, and also on the back of the cone at $b''$. In counter-revolution the two points $a''a'''$ and $b''a'''$ return to $aa'$ and $ba'$, giving $aV$ and $bV$ as the horizontal projections of the elements whose vertical projection is $a'V''$.

The important elements $VB$—$V''B'$, $VB$—$V''C'$, $VA$—$V''A'$, and $VD$—$V''A'$ are distinguished respectively as the *highest, lowest, foremost* and *hindmost*, the observer being understood to face the plane **V**.

EXAMPLES.—1°. [To become, as is very useful, familiar with revolution] revolve the base about the *horizontal trace* of its own plane, about the *vertical trace* of its own plane, and about its *vertical diameter*, in finding an element one of whose projections is given.

Ex. 2°. Perform the same operations when the base is the right-hand end of the cone.

42. Prob. XXIV., Fig. 52, shows how to find the size of the opening which a given cone of revolution in a given position would cover. Probs. XXV. and XXVI., Figs. 53 and 55, show how to find the size and direction of a cylin-

der or a cone of revolution which would cover a given elliptical opening.

After the full exhibition afforded by Pl. VI. of all the essential operations necessary in constructing the *projections* of single cones or cylinders of revolution, these bodies will, to avoid repeating these operations when merely illustrating general principles, be simply assumed in the subsequent problems.

### B—Tangencies.

43. Inspection alone, Pl. VII., Fig. 58, is sufficient to show that a *tangent plane ABC* to a cone, as *V—MTN*, or to a cylinder, touches it all along a single element as *VT*, and that it contains only that element of the surface.

Any line, as *mn*, or *pq*, in such a plane, touches the given surface only at its point of intersection with the element of contact of the plane, and is therefore a *tangent line* to the surface at that point.

44. Among all the tangent lines thus lying in the tangent plane, one, *VT*, coincides with the element of contact of the plane with the given cylinder or cone, but it is still accounted as only a particular case of the tangent line, since it is not a secant. That is, while other tangents as *mn* are tangent at *c* to any curve traced through *c* on the cone, the *element* through *c* coincides with its tangent. In other words, *the tangent to a straight line is that line itself.*

45. Since by (43) all the tangent lines at a given point of a cylinder or cone lie in the tangent plane along the element through that point, while any two intersecting lines determine a plane (Prob. IX.), the *tangent plane* to a cylinder or cone at any point is determined by *any two tangent lines at that point*, one of which by (44) may be the *element* through that point; or, derivatively, by *any lines so located as to be in the same plane with these* (27, 9°).

And as the element through a given point of contact is one of the tangent lines, *all* of which must (43) lie in the tangent plane at that point, no plane through the vertex can be considered a tangent plane unless it contains an element; while every tangent plane will contain the vertex since every element contains the vertex.

46. Comparing the preceding with the principle (Prob. IX.) that any plane is determined in the most elementary manner by three points, it is evident that the tangent plane to a cylinder or cone will be determined by any condition *equivalent to that of three determining points*, two of which are on the element of contact of the plane with the cone, and the third on any tangent at any other point on that element.

The following problems will make this principle, and the methods of applying it, more evident.

PROBLEM XXVIII.—*To construct a plane tangent to a cone on a given element.*

*In Space.*—The required plane will be determined by the given element (44) and by whatever other tangent line (43) at some point of this element is made most convenient by the position of the given cone in each case. The *traces* of the plane will then be determined by those of its determining lines.

*In Projection.*—Let the axis of the cone $VAB—V'C'D'$, Pl. VII., Fig. 57, be parallel to the ground line, and let $Vt—V't'$ be the given element of contact, one projection of which is found from the other as in Prob. XXVII.

The required tangent plane will then be conveniently determined by this element, with the tangent to any circular section, as the base $AB—C'D'$, at its intersection, as $tt'$, with this element; or, if the traces of the element are inconveniently distant, by like tangents at any two points of $Vt—V't'$ (43).

The traces of $Vt—V't'$ are $q$ and $r'$, which are therefore points of the horizontal and vertical traces respectively of the tangent plane. The tangent at $tt'$ must, being in a profile plane, be shown in revolved position in order to find its traces. Among various axes of revolution easily employed, the vertical diameter $D—C'D'$ of the base is here chosen, and the *front* half of the base is revolved to the right, carrying $tt'$ to $t''t'''$, and the vertical trace $d—dD'$ of the plane of the base to $d''—ed'''$. The tangent at $tt'$ is then shown at $d'''n'''$, a tangent at $t'''$, and $n''n'''$ and $d''d'''$ are the revolved positions, and, by counter-revolution, $n$ and $d'$ are the primitive positions of its horizontal and vertical traces respectively. Hence $nq$ is the horizontal and $r'd'$ the vertical trace of the required plane $PQP'$, whose two traces must also meet $GL$ at the same point $Q$.

EXAMPLES.—1°. Let the horizontal *trace* of the cone be given.
Ex. 2°. Substitute a cylinder for the cone.
Ex. 3°. Substitute a cylinder for the cone of Ex. 1°.
Ex. 4°. Let the axis of the cone or cylinder be perpendicular to either **H** or **V**.

PROBLEM XXIX.—*To construct a plane parallel to a given line, and tangent to a cylinder whose axis is parallel to the ground line.*

*In Space.*—All tangent planes to a cylinder being parallel to its axis, the required *planes*, of which there will be two, and hence also their *traces*, will be parallel to the ground line. Either will, therefore, be determined by its being parallel to a plane similarly situated and containing the given line, and by its containing a tangent line to the cylinder parallel to some line of this auxiliary plane.

*In Projection.*—Let $ABCD—E'F'H'K'$, Pl. VII., Fig. 60, be the given cylinder, and $ab—a'b'$ the given line. The traces of this line are $b$ and $c'$. Then $bN$ and $c'N'$, parallel to $GL$, are respectively the horizontal and vertical traces of the auxiliary plane, parallel to $GL$ and containing $ab—a'b'$. The profile plane $AQE'$ cuts from the auxiliary plane $PN—P'N'$ the line whose traces are $Y'$ and $N$, and whose position, after revolution about the vertical trace $QE'$, is $Y'N''$.

66   DESCRIPTIVE GEOMETRY.

The plane $AQE'$ also cuts from the cylinder the circle $AC-H'E'$ whose centre in the same revolution describes the horizontal arc $OO''-O'O'''$, giving the circle of radius $O'''E'''=O'E'$ as the revolved position of $AC-H'E'$. Then the tangents, $T''R''$ and $t''r''$, to this circle, and parallel to $Y''N''$, are the revolved positions of tangents to the cylinder in the two tangent planes. $R''$ is the revolved, and $R$ the primitive horizontal trace of $T''R''$, found by describing the arc $R''R$ with $Q$ as a centre. The intersection, $S'$, of $T''R''$ with the axis, is its vertical trace. Hence $RM$ and $S'M'$, parallel to $GL$, are the horizontal and vertical traces of one of the tangent planes. The other one may be similarly found.

EXAMPLES.—1°. The position of the cylinder remaining the same, let the plane be tangent on a given element.

Ex. 2°. Let the tangent plane contain a given point in space.

PROBLEM XXX.—*Through a given point in space to construct a tangent plane to a cone whose horizontal trace is given.*

*In Space.*—Any two tangents to the cone from the given point will determine the required plane. Since this plane contains the cone's vertex (45), the line from the given point to the vertex, as $PV$, Pl. VII., Fig. 59, will be one of these tangents, while, for the other, a more convenient one will be the tangent, as $AT$, from the horizontal trace, $A$, of this one, to the horizontal trace, $MTN$, of the cone. Indeed, $AT$ will be the horizontal trace of the tangent plane.

Otherwise, if any circular section of the cone be known, a tangent to it from the point where its plane cuts the tangent $PV$ at the vertex will be convenient.

*In Projection.*—Pl. VII., Fig. 61. Let $VACD-V'A'B'$ be the cone, and $aa'$ the point; then $aV-a'V'$ being a line in the tangent plane, its horizontal trace, $p$, is a point common to the horizontal traces, $PdQ$ and $PeR$, of the two planes evidently given by the solution. Since $aV—a'V'$ does not conveniently meet **V**, the vertical traces of the tangent planes are determined by the points $Q$ and $R$ with the vertical traces, $h'$ and $k'$, of the horizontal tangents, $eh—e'h'$ and $bk—b'k'$, parallel respectively to $PQ$ and $PR$, and at points $ee'$ and $bb'$ taken at pleasure on the elements of contact $Vd—V'd'$ and $Vc—V'c'$ of the planes. The vertical traces of these elements will be other points of the like traces of these planes.

EXAMPLES.—1°. Let $aa'$ be higher than the vertex.
Ex. 2°. Let it be in any other angle than the first.
Ex. 3°. Let the vertical trace of the cone be given.
Ex. 4°. Let the projections of the cone's circular base be given.

PROBLEM XXXI.—*To pass a plane parallel to a given line, and tangent to a cylinder whose axis is oblique to* **H** *and* **V**, *and one of whose traces is given.*

*In Space.*—The two simplest determining tangents (45) would be a tangent parallel to the given line, and the element through its point of contact. But the tangent plane may be determined indirectly and more simply still as follows. By Prob. IX., 4°, pass a plane through the given line and parallel to the axis of the cylinder; since all the tangent planes are parallel to the axis. Then the tangent planes, of which there will be two, will be parallel to this one, and their traces will be tangent to the like trace of the cylinder.

*In Projection.*—Pl. VII., Fig. 62. Let $ABD—A'B'$ be the horizontal trace of the cylinder, and $ab—a'b'$ the given line. $bda'$ is the plane through this line and parallel to the elements as $CE$, $A'M'$ of the cylinder, since $pq—p'q'$, in this plane, is parallel to these elements. Then $PQ$, parallel to $bd$, and tangent to $ABD$ at $T$, is the horizontal trace of one of the required planes, and $QP'$, parallel to $da'$, is its vertical trace.

The vertical trace of the element of contact $Tr—T'r'$ will also be a point of the vertical trace of the plane.

EXAMPLES.—1°. Find the other tangent plane to the above cylinder, parallel to $ab$—$a'b'$.
Ex. 2°. Let the axis of the cylinder cross any other angle than the *first*.
Ex. 3°. Let the cylinder and the given line cross different angles.

PROBLEM XXXII.—*To construct a plane parallel to a given line and tangent to a cone whose horizontal trace is known, and whose axis is oblique to both* **H** *and* **V**.

*In Space.*—The two simplest determining tangents (45) are a parallel $L$ to the given line and through the cone's vertex, and a tangent to the given trace of the cone from the like trace of $L$. Two such tangents can generally be drawn, hence the solution gives two tangent planes. If $L$ coincides with an element, there will be one plane, but none, if $L$ enters the cone.

*In Projection.*—Pl. VII., Fig. 63. Let $VAB$—$V'A'B'$ be the cone and $ab$—$a'b'$ the given line. $Vm$—$V'm'$ is the parallel to $ab$—$a'b'$ through the vertex, and its traces are $m$ and $n'$. As the tangent planes contain this line, their traces contain these traces (27, 4°), hence $smQ$ and $mrS$ are the horizontal traces of these planes, and $n'Q$ and $n'S$ are their vertical traces, which will also contain the vertical traces—not shown —of the elements of contact, $Vs$—$V's'$ and $Vr$—$V'r'$.

EXAMPLES.—1°. Let the given line cross any of the other angles.
Ex. 2°. Let the axis of the cone cross any of the other angles.
Ex. 3°. Let the projections of a circular section of the cone be given instead of its trace.
Ex. 4°. Let the construction be made for the upper nappe (32) of a cone.

47. Problems of tangent *lines* to developable surfaces, or, more definitely, to curves traced on such surfaces, might seem to properly fall here under the head of tangencies, but as such curves practically arise as the intersections of such surfaces with plane or other surfaces, such problems form a portion of those under intersections.

*Note.*—Problems XXIX.—XXXII. will enable the student here to proceed to problems of shade on cylinders and cones, and of their visible boundaries as seen from a finite distance. Thus, if the given *point* in XXX. be the place of the eye, the element of contact of the plane will be the visible limit of the convex surface ; or if the given *line* be a ray of light, that element will be the boundary between the illuminated and unilluminated part of the surface.

## C—Intersections.

48. The intersections of developable surfaces, here to be considered, are of two kinds:
 1°. Their intersections with *planes*.
 2°. Their intersections with *each other*.

The former curves, being in *planes*, are thence called *plane curves*. Otherwise, as the intersection of two surfaces *only one* of which is curved, they are also called *curves of single curvature*.

The intersection of *two curved surfaces* cannot generally also be a plane curve, and is accordingly distinguished as a *curve of double curvature*.

49. *A plane curve*, or *curve of single curvature*, is generated by a point which moves in any manner, subject only to the condition of remaining in a fixed plane. An S or an 8 are thus plane curves, as well as are circles or other simple curves.

*A curve of double curvature* is generated by a point which moves so that no more than three consecutive positions are in the same plane. Thus, let a point, $p$, starting from the position 0, take the successive consecutive positions 1, 2, 3, 4, 5.....$n$, not lying in one plane. Then, as a plane can always be passed through three points, the point $p$ will be successively in the planes 012, 123, 234......($n-2$, $n-1$, $n$).

50. The line 01 may have any one of an indefinite number of directions, giving as many possible positions of the point 1 of a curve of double curvature. For each of these positions the line 12 may have an indefinite number of directions, and

so on. Hence *there is an infinite variety of curves of double curvature.*

The same conclusion is reached from (48) by considering that there is an indefinite variety of curved surfaces, each pair of which may intersect in a great number of different ways, thus giving rise to an innumerable variety of curves, which, except in particular cases, will be of double curvature.

51. *A secant* is a line which intersects any given portion of a curve in two points, as at $ST$, Pl. VIII., Fig. 64.

Two particular cases of the secant are of special interest.

*First.* When, by turning the secant about either of its intersections as $T$, the other intersection, $S$, always continuing in the curve, comes to coincide with $T$ as at $TL$, the secant is called a *tangent*. When tangent at an infinitely distant point of the curve, it is called an *assymptote*.

*Second.* That position of the secant, as $TN$, which is perpendicular to the tangent is called a *normal*.

52. From (51) the *tangent to a curve of single curvature* may now be defined as a line lying in the plane of the curve and touching it at only one point.

The *tangent to a curve of double curvature* is likewise most simply imagined as lying in the plane of the three consecutive points of which the middle one is the point of contact. The normal at the same point lies in the same plane with the tangent.

53. *Intersections of developable surfaces with planes* are all found one by one by the general principle of Prob. XI.; that is, each point is the intersection of *some element* of the given curved surface with *some line* of the given plane which is in the *same plane with the element*.

54. The intersection of a *plane curve*, $C$, with any given plane, is found by finding, first, the intersection, $I$, of the

plane of the curve with the given plane; and, second, the points in which this line, $I$, intersects the given curve.

EXAMPLES.—1°. Find the intersection of any curve in a horizontal plane with a given oblique plane.
Ex. 2°. Let the plane of the curve be parallel to **V**.
Ex. 3°. Let it be perpendicular to either **H** or **V** only.
Ex. 4°. Let it be oblique to both **H** and **V**.

55. The intersection of a given oblique plane with a *curve of double curvature* (49) is found by the methods either of rotations (29, *Second*) or of changed planes of projection (29, *Third*), as follows:

*First.* Let $D$ be the curve, $d$ and $d'$ its projections, and $P$ the plane. Then, for example, revolve both $D$ and $P$ about the same vertical axis, $X$, till $P$ becomes perpendicular to **V**. The intersection of its vertical trace with $d_1'$, vertical projection of the curve $D_1$ (found by revolving a sufficient number of points of $D$), will then be the vertical projection, $x_1'$, of the revolved position of the required point. The horizontal projection, $x_1$, of this position will be on $d_1$, that of the like position of the curve. By counter-revolution about $X$, the real position, $xx'$, of the required point will be found on $d$ and $d'$, the projections of $D$.

*Second.* Otherwise: Assume a new plane, $V_1$, for example, perpendicular to the horizontal trace of $P$, and find $d''$ the new vertical projection of $D$ upon it, by the principle of (Prob. VI.) The trace of $P$ upon $V_1$ will there intersect $d''$ at a point $x''$, auxiliary vertical projection of the required point. From this $x$, on $d$, and thence $x'$, on $d'$, can at once be found.

EXAMPLE.—Find the intersection of the plane $PQP'$, Pl. VI., Fig. A, with the curve of double curvature $dd'$.

56. The following problems will sufficiently illustrate the construction of the intersections of planes with cylinders and cones, being those marked by a star in the following table of the various combinations; where a right plane or cone means one that is perpendicular either to **H** or **V**.

## PLANE INTERSECTIONS.

|  |  | Plane. | |
|---|---|---|---|
|  |  | Right. | Oblique. |
| Cylinder.................... | Right. | 1 | 2 |
|  | Obl. | 3* | 4* |
| Cone........................ | Right. | 5 | 6* |
|  | Obl. | 7 | 8* |

PROBLEM XXXIII.—*To find the intersection of a plane perpendicular to* **V**, *with a cylinder whose axis is oblique to both planes of projection, and whose horizontal trace is given.*

*In Space.*—The plane being perpendicular to **V**, so much of its vertical trace as lies between the extreme elements of the cylinder is (27, 8°) the vertical projection of the entire curve; any point of which is therefore the intersection of this projection with the vertical projection of an element of the cylinder. The horizontal projection of the point will then be on that of the same element, whose horizontal trace is on that of the cylinder.

*In Projection.*—Pl. VIII., Fig. 67. Let $ACD$ be the horizontal trace of a cylinder of revolution, whose projections by (Prob. XXV.) are $ABDbd$ and $E'C'c'e'$; and let $PQP'$ be the given plane. Then $c'e'$ is the vertical projection of the required intersection. Then, taking the extreme elements, and other convenient ones, as $B'b'$, whose horizontal projection is $Bb$, and $Ff'$, whose horizontal projection is $Ff$, project the point $b'$, intersection of $P'Q$ and $B'b'$, upon $Bb$, giving $b$, and $f'$, intersection of $Ff'$ and $P'Q$, upon $Ff$, giving $f$. Having in this way found a sufficient number of points of the horizontal projection of the curve, connect them, as shown, observing to make the upper half of the curve full, since it is visible.

This part is on elements whose horizontal traces are on the half *DEB* of that of the cylinder.

EXAMPLES.—1°. The cylinder remaining as in Fig. 67, let the plane *PQP'* be vertical.
Ex. 2°. Solve case (1) of the table. [See my Elem. Proj. Drawing.]
Ex. 3°. Solve case (2) of the table. [See Fig. 65.]
Ex. 4°. Let the cylinder be given by its axis and diameter.
Ex. 5°. Substitute a cone, oblique both to H and V, for the cylinder, *PQP'* remaining as now.

## PROBLEM XXXIV.—*To find the intersection of a plane with a cylinder, both being oblique to both planes of projection.*

*In Space.*—Any auxiliary secant plane, parallel to the axis of the cylinder, will cut from the surface of the cylinder *two elements*, and from the given plane a *straight line*. Where this line meets those elements will be two points of the required curve.

The auxiliary plane is most conveniently taken perpendicular to a plane of projection.

The solution is thus seen to be the same as that of Prob. XI.

*In Projection.*—Pl. VIII., Fig. 68. Let *ADB* be the horizontal trace of a cylinder of revolution, *HIhi—A'B'R'*, and let *PQP'* be the given plane. Let the auxiliary planes be vertical; they will also be parallel, since the elements of the cylinder are so, and their intersections with the given plane will therefore be parallel. *MNP'*, where *MN* is parallel to *Ii*, is parallel to the auxiliary planes, hence its intersection, *MN—m'P'* with *PQP'*, found by Prob. X., is parallel to those of *PQP'* with these planes. This done, any secant auxiliary plane, as *FE*, contains the elements *Ff—F'f'* and *Ee—E'e'* of the cylinder, and cuts from the given plane *PQP'* the line *Ek—c'k'*, parallel to *MN—m'P'*. Hence

$ff'$ and $ee'$, where those elements intersect this line, are the points where these elements meet $PQP'$, and hence are points of the required intersection.

Other points being similarly found, and joined in each projection, the required curve $hde-h'd'e'$ will be found.

In the figure, the vertical projection of each point is found first, because the auxiliary planes are vertical. As in Prob. XI., were these planes perpendicular to **V**, the horizontal projection of each point would be found first.

The visible half in horizontal projection is the upper one, whose points are on elements having their horizontal traces in the half $HBI$ of the trace of the cylinder. The front half is visible in vertical projection, and crosses the elements whose horizontal traces are on the half $AHB$ of the trace of the cylinder.

EXAMPLES.—1°. Take the auxiliary planes perpendicular to **V**.
Ex. 2°. Let the given cutting plane have the kind of position (Prob. IV.) indicated in Fig. 71.
Ex. 3°. Let the axis of the cylinder cross the first angle.
Ex. 4°. Let the given trace of the cylinder be on **V**.
Ex. 5°. Let the cylinder be given by its axis and diameter.

**PROBLEM XXXV.**—*To find the intersection of an oblique plane with a cone whose axis is vertical.*

*In Space.*—Recollecting that a circle, if parallel to a plane of projection, is as convenient an auxiliary as a straight line, the *auxiliary* planes in this problem may have any position in which they will intersect the given cone, either in straight lines or circles. Then the points in which the line cut from the conic surface by any auxiliary plane meets the line cut from the given cutting plane by the same auxiliary plane, will, being thus common both to the cone and the given plane, be points of their required intersection.

*In Projection.*—Pl. VIII., Fig. 69. Let $VABK-V'A'B'$ be the cone, and $PQP'$ the cutting plane, so taken that its traces coincide. As will be seen, this position of $PQP'$ only requires care in distinguishing its traces by their uses in the successive steps of the solution.

1°. *The points on the extreme elements, $V''A'$ and $V''B'$.*—To find these, take the auxiliary plane $ABP$ which contains these elements, and, remembering that it is parallel to **V**, its intersection, $PV-P'r'$, with $PQP'$, is parallel to the vertical trace $QP'$. Then $n'n$ and $s's$, the intersections of $p'r'-PV$ with $V''B'-VB$ and $V''A'-VA$, are the points required. As $p'r'$ is in the cutting plane $PQP'$, and also in the same plane with the cone's axis, it intersects this axis at $r'$, which is therefore the intersection of $PQP'$ with the cone's axis.

2°. *The highest and lowest points.*—These are evidently on the elements lying in a plane perpendicular to the given plane, and containing the axis of the cone. $VC$, perpendicular to $PQ$, is the horizontal trace of such a plane, and it contains the elements $VL$ and $VH$. Revolving the plane $VC$ about the cone's axis till it is parallel to **V**, these elements appear in vertical projection at $V''B'$ and $V''A'$; and the *line $VC$*, intersection of the *plane $VC$* with $PQP'$, appears at $C''V-C'''r'$, since the cutting plane, as shown in (1°), intersects the cone's axis at $r'$. Hence $l'''$ and $h'''$, intersections of $C'''r'$ with $V''B'$ and $V''A'$, are the revolved positions of the required highest and lowest points. By counter-revolution, they return to their real positions, $l'$ and $h'$, on $C'r'$, the primitive position of $C'''r'$. Their horizontal projections, $l$ and $h$, may be found by projecting $l'$ upon $VL$ and $h'$ upon $VH$, or by projecting $l'''$ and $h'''$ at $l''$ and $h''$, and thence revolving these, as shown, by arcs, with $V$ as a centre, to $l$ and $h$.

3°. *Points on the foremost and hindmost elements.*—These elements are those whose horizontal projections are $VI$ and $VK$, and which are both vertically projected in $V'o'$. Their projections thus coincide with the traces of the auxiliary plane $ImV''$ containing them, and with $om-o'm'$, the intersection of this plane with the given plane $PQP'$. Hence, revolving their plane about the cone's axis, till parallel to **V**, they will appear at $VB-V''B'$ and $VA-V''A'$, and $om-o'm'$ at $o''m''-o'''m'''$. Hence $q'''q''$ and $k'''k''$ are the revolved positions, and $qq'$ and $kk'$ the primitive positions of the points on the foremost and hindmost elements.

4°. *Points on any other elements.* (See Fig. 70.)—Assuming $V'B'$, for example, at pleasure, either as the vertical projection of two elements, or as the vertical trace of their plane, $VA$ and $VB$ are the horizontal projections of the elements, and $PB'$, perpendicular to the ground line $D'Q$, is the horizontal trace of their plane. Then $PQP'$ being the given plane whose intersection with the cone $VABC-V'C'D'$ is required, $Pm-B'm'$ is its intersection with the auxiliary plane $PB'V'$, and $aa'$ and $bb'$, where this intersection meets the assumed elements $VA-V'B'$ and $VB-V'B'$, are those points of the intersection of the given cone with the given plane, which are on these elements.

5°. *Points found by horizontal auxiliary planes.*—Let $P'R'$, Fig. 69, be the vertical trace of such a plane. It cuts from the cone the circle of radius $d'v'$, whose horizontal projection has $Vd$ for its radius, and from the given plane $PQP'$ the horizontal line $P'R'-fe$, where $fe$ is parallel to the horizontal trace $PQ$. Then $ee'$, and another point, very near $nn'$, intersections of $fe-P'R'$ with the circle $Vd-v'd'$, are two points of the required intersection of $PQP'$ with the given cone.

EXAMPLES.—1°. Complete the construction of the curve (Fig. 70) of which $aa'$ and $bb'$ are two points.

Ex. 2°. Find the intersection of a right cone with a right plane (see Fig. 66).

Ex. 3°. In Ex. 2 let the plane perpendicular to **V** be parallel to an element of the cone.

Ex. 4°. Replace the cone by a pyramid.

Ex. 5°. Replace the cone by any right or oblique prism.

PROBLEM XXXVI.—*To find the intersection of a plane and a cone, both of which are oblique to both planes of projection.*

*In Space.*—As before, any point of the required intersection will be where an element of the cone meets a line of the given plane, both being in the same auxiliary plane. But as it is most convenient to take the auxiliary planes in some uniform manner, let them all contain the cone's vertex and be perpendicular either to **H** or **V**; they will then all contain a line having a like position, hence their traces on the given plane will (27, 4°) all pass through the point where this line pierces that plane.

*In Projection.*—Pl. VIII., Fig. 71. Let $VBD-V''A'C'$ be the given cone, and $PQP'$ the given plane. Let the system of auxiliary planes through $VV'$ be perpendicular to **H**, they will therefore intersect each other in the vertical line $V-s'V'$. By (Prob. XI., $b$, 1°) this line pierces $PQP'$ at the point $V,s'$, found by taking any line $Vx-x'y'$, containing $V$, and in this plane, as shown by its traces, $x$ and $y'$, in those of the plane, either $xx'$ or $yy'$ being assumed. The point $V,s'$ is then common to all the lines cut from $PQP'$ by the vertical auxiliary planes through the cone's vertex. $ECV$ is the horizontal trace of one of these planes; it cuts from the cone the elements $EV-E'V'$ and $CV-C'V'$, and from $PQP'$ the line $ErV-s'r'$, which meets these elements at $e'e'$ and $c'c'$, which are, therefore, two points of the required intersection. Others, as many as necessary, always taking those on the extreme elements of both projections, can be found and connected as shown.

The visible points of the curve, as in Fig. 68, are on the visible elements of the cone.

**57.** *Intersections of developable surfaces with each other.*—These are all found on the general principle that each point of the intersection is where an element of one of the surfaces meets one from the other surface, the two elements being in the same plane. They are sufficiently exemplified by the following problems. If the intersecting surfaces have a common axis, their intersection will be the circle perpendicular to that axis, generated by the point of intersection of the elements in the common meridian plane (33).

PROBLEM XXXVII.—*To find the intersection of two cones, the axis of each being oblique both to* **H** *and* **V**.

*In Space.*—Any plane *AIV*, Pl. IX., Fig. 72, containing elements of a cone contains its vertex. But any number of planes can be passed through two given points, hence planes can be passed through both the given vertices, *V* and *v*, each of which will, therefore, contain elements of both cones. Such planes will all contain the line *Vv* joining the two vertices, and hence their horizontal traces, as *IA* and *IB*, will contain *I*, that of this line (27, 4°), *which is, therefore, the first thing to be found.* These traces will then intersect the like traces of the cones in the traces as *A* and 4 of the elements contained in those planes. Then, *where the elements, as Av and Cv, of one cone, contained in any one of the planes, as AI, meet those* 1*V and* 4*V of the other cone, and contained in the same plane, will be points, as a, $a_1$, c, etc., of the required intersection.*

This solution is general for every case of pairs of bodies, each of which has a vertex.

*In Projection.*—Pl. IX., Fig. 78.

1°. *Selection of planes and construction of points.*—Let $VdLb-V''c'f'$ and $vADG-v'C'H'$ be the given cones. Following the solution in space, *I* is the horizontal trace of the line $Vv-V'v'$ joining the given vertices. Hence, all lines through *I* and secants of the cones' bases, will be the horizontal traces of planes, each containing elements of both cones. Such planes will be limited, as at *IbB* and *IfE*, by those which, being tangent to one cone, cut the other, or may be tangent to both.

Planes between these limits are chosen, not at random, but so as to contain the extreme elements of both cones relative to both projections, as *IC* contains an extreme element *V'C'* of the vertical projection of cone *vv'*, and as *IG* contains one of its extreme elements in horizontal projection. More planes than are shown for illustration were therefore necessary in accurately finding the curve. Some of these planes may often exactly or sensibly contain two or more extreme elements.

Take now, for example, the plane $IC$. It cuts from the cone $VV'$ the elements $hV-h'V'$ and $cV-c'V'$, and from the cone $vv'$ the elements $Hv-H'v'$ and $Cv-C'v'$. The latter intersect the former in four points, as found by following them up, two by two, from the bases, at 2,2′, 12,12′, 6,6′, and 8,8′.

When, as at $IE$, a plane is tangent to one of the cones, it contains but one element of that cone, and therefore gives but two points, as 4,4′, and 10,10′, of the required intersection.

Any desired number of points may be similarly found.

2°. *Connection of Points.*—*Rule.* Begin at any point, and proceed from plane to plane, joining the points, till a tangent plane is reached. There the curve will be tangent to the elements cut from one cone by the plane which is tangent to the other. Thence the curve recrosses the elements already traversed, on the body cut by this tangent plane, through points on new elements of the other body.

Thus proceed till another tangent plane is reached, and thence in like manner to the point of beginning.

This is evident from Fig. 72, for as there can be no auxiliary plane exterior to the tangent one $IB$, that limits the curve, so that in passing from $a_1$, in plane $IA$ to $b$, its contact with $V2$ in plane $IB$, it must thence return to plane $IA$, but at a point $c$, where a new element, $Cv$, of cone $v$, crosses the same element $V1$ of cone $V$, that contains $a$.

*Illustration.*—Thus, in Fig. 78, beginning anywhere, as at 1,1′, where the curve is tangent to the element $Bv-B'v'$ of cone $vv'$, sketch the curve through 2,2′, 3,3′, and other intermediate points not shown, to 4,4′, where it is tangent to the element $Ve-V'e'$ cut from cone $VV'$, by the plane $IE$, tangent to cone $vv'$. Thence proceed through 5,5′, 6,6′, to 7,7′, where it is again in the tangent plane $IB$, but tangent to the other element, $Kv-K'v'$, cut by that plane from the cone $vv'$. Thence through 8,8′, 9,9′, the curve, once more in the tangent plane *IE*, and tangent at 10,10' to the element $Vf-V'f'$, returns to 1,1', the starting-point chosen.

3°. *Number and disposition of curves.*—If both of the tangent planes to one body should cut the other, there would be two curves, as in Fig. 75.

If one of the auxiliary planes should be tangent to both bodies, it would contain but one element of each, and would therefore give but one point, at which the curve would cross itself.

Should two of the auxiliary planes each be tangent to both cones, each would give one point, making two points at which the curve would cross itself, and the curve would be found reduced to two plane curves which would therefore generally be two ellipses. (Theor. IV.)

4°. *Visibility of the curve.*—In order that a point of the curve should be visible, in either projection, it must be at the intersection of elements both of which are visible in that projection. Thus the point 11,11' is visible in horizontal projection, and so is the point 4,4' in vertical projection. Having determined such a point, the curve is visible in both directions from it, until it reaches the nearest extreme elements found in following the curve. Thus from 11, the curve is visible to 9 on *vG*, and to the point between 12 and 1 on *Vm*.

5°. *Visibility of extreme elements.*—This is determined by that of the elements which intersect the extreme elements. Thus *Gv* intersects the just invisible element *dV* at 5, where it enters cone *VV'*, and emerges at 9 on the visible element *Vg*, and hence is visible from 9 to *v*.

Likewise, the extreme element *C'v'* intersects the two invisible ones *c'V'* and *h'V'*, and hence is invisible between the extreme elements *V'n'* and *V'f'* of cone *VV'*.

EXAMPLES.—1°. Construct the problem so as to give two curves. [To certainly do this, assume one cone as *vv'*, and its tangent planes from *I*, also assumed. Then place the vertex *VV'* on *Iv—I'v'* and the base of cone *VV'* so that both of these tangent planes shall cut it.]

Ex. 2°. Take the given cones so as to make a more visible curve. [For example, so that the chord *ef* shall be much nearer *L*.]

DESCRIPTIVE GEOMETRY. 81

Ex. 3°. Placing the ground line at right angles to its present position, make the total figure higher than it is wide, and so as to fill the plate.

Ex. 4°. Let one base be within the other.

Ex. 5°. Let $II'$ be to the left of the cones and further forward from the ground line than the bases.

58. By removing the vertex of one of the cones to infinity on its axis, it becomes a cylinder as sketched in Pl. IX., Fig. 74.

The line $Vv$ then becomes a parallel to the axis of the cylinder, through the vertex of the cone, as shown at $VI$, Fig. 74. With this modification of the initial step, only, by making all the auxiliary planes contain $IVv$, the solution remains the same in all particulars as that of Prob. XXXVII.

59. By similarly removing the vertices of both cones, they become two cylinders as shown in Pl. IX., Fig. 73.

In this case, by (Prob. IX., 4°) we pass *an initial auxiliary plane*, as $Op$, through the axis, as $Om—O'm'$, of one cylinder, and parallel to that of the other one. The other auxiliary planes are then parallel to $Op$, as indicated by their parallel traces, and the solution thenceforward is in all respects the same as that of Prob. XXXVII.

PROBLEM XXXVIII.—*To find the intersection of two cylinders whose axes intersect at right angles, one of them being vertical, and the other parallel to the ground line.*

*In Space.*—All that is peculiar to this problem, as compared with (59), arises from the position of the horizontal cylinder. The auxiliary planes being placed as there described, the projections of elements of that cylinder may be found as in Prob. XXVII.

*In Projection.*—Pl. IX., Fig. 75. Let $hr$ be assumed as the horizontal trace of an auxiliary plane. It cuts from the vertical cylinder two elements, horizontally projected at $g$ and $b$, and from the horizontal cylinder two elements whose horizontal projection is $hr$. The vertical projections of these are here shown by revolving the plane of the base $po$ about its horizontal diameter, $po$—$p'$, till it becomes horizontal, when the circle $pqo$ will be its horizontal projection, and the two points $r$ will appear on it as at $r''$—here supposed to be the upper one—and a point, not shown, equally to the left of $r$. The vertical projection of $r''$ is $r'''$, and serves alike for finding by counter-revolution, as shown, $r'$, the vertical projection of $r$, and $k'$, the vertical projection of the lower point projected in $r$.

Thence, as is evident on inspection, the auxiliary plane $hr$ determines the four points, $bu'$, $bb'$, $gt'$, $gg'$.

The construction of the highest points, as $nd'$; foremost, as $aa'$; lowest, as $nf'$; and hindmost, as $la'$, is obvious; as is the symmetry of the curves relative to the vertical plane $DO$, and horizontal plane $A'O'$.

EXAMPLES.—1°. Construct the intersection of a cylinder and a cone for the case where there are two curves.

Ex. 2°. Construct the intersection of a cylinder and a cone for the case where there is but one curve.

Ex. 3°. Let the axis of the *cylinder* be vertical, and that of the cone parallel only to **H**.

Ex. 4°. In (3) let the axis of the cone be parallel to the ground line.

Ex. 5°. Let the axis of the *cone* be vertical, and vary the position of the cylinder as that of the cone is in (3) and (4).

Ex. 6°. Further vary (3), (4), and (5) by making either body penetrate the other, giving two curves.

Ex. 7°. Construct the intersection of two cylinders, placed as in Fig. 73.

Ex. 8°. Place the two cylinders, so that there shall be two curves.

Ex. 9°. Vary Fig. 75 by making the axis of the horizontal cylinder there shown, oblique to **H**, or to **V**, or to both planes.

Ex. 10°. Vary Fig. 75 by revolving the plane of the base $po$—$p'c'$, either about its horizontal trace, or its vertical trace, or about the diameter $c$—$v'c'$.

Ex. 11°. In Fig. 75, let the axes, $DO$—$A'O'$ and $O$—$m'Y$, not intersect.

## PROBLEM XXXIX.—*To construct the tangent line at a given point of any previous plane, or double-curved intersection.*

*In Space.*—1°. The tangent line to any plane curve being in the plane of the curve (52), and the tangent line to a surface at any point being in the tangent plane at that point (43), the required tangent line to any plane intersection will be the intersection of these two planes.

2°. By the second principle just given, the tangent line at a given point of the curve of intersection of two developable surfaces will be the intersection of two planes, each of which is tangent to one of the surfaces along the element containing the given point; as illustrated in Pl. IX., Fig. 74, by *Qt*, the tangent at *t*, and the intersection of the plane *EQt*, tangent along *Et* to the cylinder, with the plane *FQt*, tangent along *Ft* to the cone.

*In Projection.*—Pl. VIII. By the above principles (1°), in Fig. 67 the tangent at *ff'* to the curve *bcd—c'e'*, is *Pf—Qf'*, which is the intersection of the plane *PQP'* of the curve, and the tangent plane *PF* along the element *Ff—F'f'* containing the given point of contact *ff'*. The point of contact being known, one trace of the tangent plane is sufficient.

Likewise in Fig. 68, the tangent *Ke—K'e'*, at the given point *ee'* of the curve *fhe—f'h'e'*, is found as the intersection of the plane *PQP'* of the curve, with the tangent plane *KE* along the element of the given cylinder containing *ee'*. *K* being the intersection of the horizontal traces, *KE* and *PQ*, its vertical projection *K'* is on the ground line.

Also in Fig. 69, the tangent *Tn—B'n'* at *nn'* to the curve *hkn—h'k'n'*, is the intersection of the plane *PQP'* of the curve, with the tangent plane *BB'V'* along the element *BV—B'V'* containing *nn'*. Tangents at *aa'* and *bb'* in Fig. 70, and the tangent *Kf—K'f'* in Fig. 71, are similarly found.

2°. Pl. IX., Fig. 78. The tangent *M*8—*M'*8', to the intersection of the two cones, at the point 8,8', is the intersection of the tangent planes, *HM*, to the cone *vv'*, along the element *Hv—H'v'*, and *hM* to the cone *VV'*, along the element *hV—h'V'*.

EXAMPLES.—1°. Construct the tangent at any other given point in each of the foregoing figures.

Ex. 2°. Construct the tangent line at the point *bb'*, Fig. 75.

## D—Development.

**PROBLEM XL.**—*To find the true form and size of the intersection of any cylinder or cone by a plane, with the tangent to the revolved curve.*

*In Space.*—The given plane is revolved so that it shall either coincide with or be parallel to a plane of projection, when the given curve will be seen in its true form and size on that plane. (See Probs. XII. and XV.)

*In Projection.*—Pl. VIII. 1°. In Fig. 67, the given plane $PQP'$ is perpendicular to **V**, hence if revolved into **V** about $P'Q$ as an axis, any point, as $ff'$, will be found as at $f'''$ (not shown) on a perpendicular to $P'Q$ at $f'$, and at a distance from $P'Q$ equal to $fg$. Similarly finding $P'''$, revolved position of $P$, the revolved position of the tangent $Pf$—$Qf'$ will be $P'''f'''$.

Or, revolving $PQP'$ about $PQ$ into **H**, $ff'$ will appear at $f''$, by making $f''h = f'Q$, and, $P$ remaining fixed in the axis, $Pf''$ is the revolved tangent.

2°. In Fig. 68, $ee'$, for example, revolved into **V**, about the vertical trace $P'Q$ of the plane $PQP'$ of the curve $ehf$—$e'h'f'$, will fall at $e''$, at a distance from $P'Q$ equal to the hypothenuse constructed on $e'p$ and $eq$ as sides. (Prob. XII.) The tangent at $ce'$ pierces **V** in $P'Q$ at $g'$, hence $g'e''$ is its revolved position.

3°. In Fig. 71, we may proceed just as in the preceding cases, or make use of the system of lines meeting at the point $V, s'$. Revolving the curve $ebf$—$e'b'f'$ about $PQ$ into **H**, this point $V, s'$ in the plane $PQP'$ of the curve appears at $s''$, where $s''m$ is the hypothenuse found from $Vm$ and $s'o$ as sides. Then $s''$ is common to the revolved positions of the system of lines $Vn$—$s'n'$, etc., in $PQP'$, which will therefore appear at $s''n$, etc. The curve and the point $V, s'$ being on opposite sides of the axis $PQ$, will so appear after revolution. Thus $ff'$ appears at $f''$, the intersection of $s''n$, produced, with $fi$, perpendicular to $PQ$ and the horizontal projection of the arc in which $ff'$ revolves about $PQ$. The horizontal trace, $K$, of the tangent at $ff'$, being in the axis, remains fixed, hence $Kf''$ (not shown) is its revolved position.

EXAMPLES.—1°. In Fig. 67, let the axis be in $PQP'$ and *parallel* to $P'Q$ or $PQ$.

Ex. 2°. In Fig. 68, revolve the curve about $PQ$, or a parallel to it, in $PQP'$, till in or parallel to **H**.

Ex. 3°. Show the true size of the curve in Fig. 69.

Ex. 4°. In Fig. 71, find the true size of the curve as in preceding examples; also by using the point $V, s'$, but by revolution about $P'Q$ as an axis.

60. Any two elements, either of a cylinder or cone, lie in the same plane; then if $a, b, c, d\ldots$ be consecutive elements, any one, as $a$, may be revolved about the consecutive one, $b$, till it falls into the plane of the consecutive elements $b$ and $c$; next the plane of $a$, $b$, and $c$ may be revolved about $c$ into the plane of $c$ and $d$; and so on till all the elements are brought into one plane. The surface is then said to be developed upon a plane, and as the *relative position* of consecutive elements is unchanged, the curved surface is merely transformed into an equivalent plane area, but not rent or destroyed. Hence, also, the relative positions of lines having a fixed relation to these elements, as any curve on the surface and its tangent, will be unchanged by development.

61. *Illustration.*—A material cone being placed with one of its elements in a flat surface of paper, let it be rolled without slipping until the same element again coincides with the paper. The portion of paper thus traversed, if cut out, will be found to be of the form of a sector of a circle, and, if wrapped upon the cone, will exactly cover its convex surface. It is therefore the development or pattern of that surface.

PROBLEM XLI.—*To develop the convex surface of a cylinder, together with its intersection with a plane, or with another cylinder or a cone, and any tangent line to the curve.*

*In Space.*—The development of the convex surface of a cylinder limited by circular bases will by (60) be a rectangle, whose length is the circumference of the base of the cylinder, and whose altitude is the same as that of the cylinder. Whatever the position of the cylinder, the length of any portion of any element given by its projections can be found by Prob. XII.

The developed tangent is readily found by means of its true length and direction. Thus, it is a hypothenuse, of which the adjacent sides are a *perpendicular from any point of it to the element containing the point of contact*, and the *portion of this element included between the tangent and this perpendicular*.

*In Projection.*—Pl. IX., Figs. 75-77. In Fig. 76, which is the development of the vertical cylinder in Fig. 75, $mm_1$ equals the circumference of the circle whose centre is $O$, and $mY = m'Y$ (Fig. 75). Then in the respective figures, 76 and 75, $m_1x = ma$, and $a_1x = xa'$; $xy = ab$, and $yb = yb'$, etc., giving the development of the curve of intersection $anl—f'a'd'$.

In Fig. 77, which is the development of the horizontal cylinder in Fig. 75, $Dc = Dc$ in Fig. 75, and $DD_1 =$ the circumference of the base, $po—v'c'$, of the horizontal cylinder. Then, in the same figures respectively, $cr = qr''$, and $rb = r'b'$; $rp = r''p$ and $pa = p'a'$; $pk = pr''$ and $ku = k'u'$; $kv = r''q$ and $vf = v'f'$, etc., and $dbauf$ is the development of the front half, $nba—d'b'a'u'f'$, of the intersection, considered as on the horizontal cylinder. The tangents at $d$, $a$, $f$, $a_1$ and $d_1$ are evidently parallel to $cc_1$.

EXAMPLES.—1°. Construct the tangent at $b$ (Figs. 76 and 77), development of that at $bb'$ (Fig. 75).

Ex. 2°. Develop the cylinders and intersections shown in Pl. VIII., Figs. 67, 68 ; supposing them true cylinders of revolution (42).

PROBLEM XLII.—*To develop the convex surface of a cone of revolution with any curve on its surface, and a tangent to the curve.*

*In Space.*—The development of a cone of revolution will by (61) be the sector of a circle, whose radius is the slant height of the cone, and whose arc equals the circumference of the cone's base.

The developed tangent is found as in the last problem.

*In Projection.*—Pl. IX., Figs. 79–81.  1°. Fig. 79 is the development of Fig. 70 on Pl. VIII. Hence $VE$ (Fig. 79) $= V'D'$ (Fig. 70), and the arc $EE_1$ equals the circumference of the base $ABC$ of Fig. 70. The cone is here supposed to be placed with the element $VE-V'E'$ (Fig. 70), tangent to the plane of the paper at $VE$ in Fig. 79, and then rolled until the same element again reaches the paper at $VE_1$, giving on the respective figures, 79 and 70, $EB = EB$, $BD = BD$, $Ad = Ad$, etc. The developments, as $a$ and $b$, of points, as $aa'$ and $bb'$, of the intersection of the cone with the plane $PQP'$, are found by making $Va$, etc., equal to the true length of $Va-V'a'$. This is best found by revolving $Va-V'a'$ about the cone's axis till it is parallel to **V**, when, as the cone is one of revolution, with its axis vertical, $Va-V'a'$ will fall upon $V'D'$ as at $V'a''$, by drawing $a'a''$ parallel to $GL$ (7), to represent the arc in which $aa'$ revolves to $a''$. The tangent $be$ is then found by making $Be$ (Fig. 79) $= Be$ (Fig. 70), and drawing $be$. The tangent $aM$ is similarly found.

2°. Fig. 81 sufficiently represents the method of proceeding to develop a cone of revolution, placed as in Pl. VI., Fig. 52, and represented on Pl. IX. by Fig. 80, indicating the projection $V''MR$ (Fig. 52) on a vertical plane through the cone's axis.

Here $V''C$, which bisects the angle $MV''R$, is the axis of the cone, and $NCr$ perpendicular to it is therefore a circular section of the cone, half shown as such at $Ncr$, by revolution about $Nr$ into the plane of the paper. Then in the development (Fig. 81), $V''N = V''N$ in Fig. 80; $Na$, $ac$, etc., are equal to the like arcs on Fig. 80; and if $p$ (Fig. 80) is any point upon the cone's surface, its true distance from $V''$ is $V''p_1$, found by drawing $pp_1$ parallel to $Nr$, as in the previous case; hence $V''p_1$ is laid off on the element $V''a$, development of $V''A$, at $V''p$.

The elliptical base $MR$ is similarly developed. Thus

$V''R$ (Fig. 81) equals $V''R$ on Fig. 80, $V''K_1 = V''K_1$ (Fig. 80), $V''M$ (Fig. 80) would be laid off on both $VN$ and $VN_1$, produced.

EXAMPLES.—1°. Make the complete development, just indicated, of Pl. VI., Fig. 52.

Ex. 2°. Develop each of the cones in Pl. IX., Fig. 78, supposing them first to have been accurately constructed as cones of revolution.

Ex. 3°. Develop the figure of Ex. 11°, Prob. XXXVIII.

## CHAPTER III.

### WARPED SURFACES OF REVOLUTION.

62. A WARPED surface is generated by a straight line which moves so that its consecutive positions are not in the same plane.

If there be a warped surface of revolution, it must be generated by the revolution of a straight line around another straight line not in the same plane with it; for the only other positions which the revolving line can have with respect to the axis are, parallel to it, and intersecting it. In the former case a cylinder will be generated, and in the latter a cone, but both of these surfaces are developable (60).

THEOREM V.—*The surface generated by the revolution of a straight line around another straight line not in the same plane with it, is warped.*

This theorem is proved by showing that the consecutive positions of the revolving line, that is, the consecutive elements of the surface, are not in the same plane.

See then Pl. X., Fig. 82, in which the vertical line $O$—$O'O''$ is the axis, $ab$—$a'b'$ the generatrix, and the shortest or perpendicular distance, $Oc$—$c'$, between the two lines is horizontal, and sweeps over the circular area of radius $Oc$, called the *circle of the gorge*.

Then, first, since no point of $ab$—$a'b'$ is on the axis, every point of it moves, hence consecutive positions have no common point, or *do not intersect*.

But, second, as $ab$—$a'b'$ remains perpendicular to $Oc$—$c'$ while both revolve, the horizontal projections of the successive positions of $ab$—$a'b'$ are tangent to the circle of radius $Oc$, and hence are not parallel, and therefore are also *not parallel* in space.

Thus the consecutive elements of the surface are not in the same plane, and therefore the surface composed of them is warped.

THEOREM VI.—*The warped surface of revolution has two sets of elements, such that at every point of the surface two elements can be drawn, one of each set.*

This theorem is evident (see Pl. X., Fig. 85) from the symmetry of the two lines $ab$—$a'b'$ and $ab$—$a''b''$, relative to the plane containing the axis $O$—$h''h'$ and the common perpendicular, $Oq$—$q'$, to it and to both of these lines. For by reason of this symmetry, any two points, as $aa'$ and $bb''$, one on each element, and in the same plane perpendicular to the axis, are equidistant from the axis, and will therefore describe the same horizontal circle; thus the series of circles described by all the points of $ab$—$a'b'$ is identical with that described by all the points of $ab$—$a''b''$. Hence the surfaces generated by these lines are identical.

The two sets of elements are distinguished as those of the *first and second generations.*

Again: from any point, as $n$ on $ab$, draw a tangent $nr$. Then $nr = nq$. But each of these represents two lines, one above and one below the plane $Oqr$, and all equally inclined to it. Hence, taking, for example, the pair that *ascends* from $q$ and $r$ to $n$, they, being equal, because equal in horizontal projection and in inclination to H, really intersect at a point whose projection is $n$. But they are of different generations, for $rn$, ascending from $r$, is an element of the generation to which $ab$—$a'b'$ belongs, while $qn$, ascending from $q$, is an element of the generation to which $ab$—$a''b''$ belongs. Similar reasoning applies to $n$ when taken anywhere on any position, either of $ab$—$a'b'$ or of $ab$—$a''b''$. Hence, at every point of the surface two *elements may be drawn, one of each generation.*

63. It follows, from the last theorem, and considering all the consecutive positions of *n* on any one element, that, in the warped surface of revolution, *any element of one generation intersects all those of the other generation.*

64. A line is fixed in position when it intersects three given straight lines not in the same plane, and contains a given point on one of them. For let *A*, *B*, and *C* be the three given lines, and *a* the given point on *A*. The plane containing the point *a* and line *B* is then fixed, and not containing the line *C* will intersect it at some point *c*. The line *ac* will then be determined by the two points *a* and *c* and will intersect the three given lines.

### A—Projections.

PROBLEM XLIII.—*To construct the projections of the warped surface of revolution.*

*In Space*—I. *General Construction.* Since each element of one generation intersects *all* those of the other (63) it will intersect *any three* of them. If then we have any three lines whose perpendicular distances from a fourth, taken as an axis, are equal and in the same plane; and which are equally inclined in the same sense to that plane, they will be elements of one generation of the required surface, and will therefore be *directrices* from which, by (Theor. VI.), any number of elements of the other generation can be found.

II. *Special Constructions.*—In these, having given the axis as vertical, and an element parallel to **V**, we know at once the radius of the gorge, and the inclination of all the elements to **H**, and thence can find the horizontal trace of the surface, and any number of elements, the horizontal projections of the latter being (Theor. V.) tangent to the like projection of the gorge.

*In Projection.*—Pl. X., Fig. 85. Let the surface be given by its axis $O$—$h''h'$, and the element $ab$—$a'b'$, parallel to **V**. Then at once $ab$—$a''b''$ is the element of the other generation at $qq'$. Every tangent to the circle $Oq$ is likewise the horizontal projection of two elements, one of each generation. Thus, $hk$ is the horizontal projection of the pair of elements whose vertical projections are $h'k'$ and $h''k''$.

See also $ef$—$e'f'$ and $ef$—$e''f''$, which intersect at $oo'$; $ef$—$e''f''$ and $eg$—$e''f'''$, which intersect at $ee''$, etc.

*Visibility.*—In horizontal projection, the inner side of that part of the surface which is above the gorge is visible, and in vertical projection, the part which is in front of the meridian plane $ey$ is visible. But the symmetry of the elements, relative to both of these limits, makes every line, which represents elements, visible throughout. Thus $hp$ is visible as the horizontal projection of $h''p'$ and $kp$ as that of $k'p'$. Likewise $h''s''$ is visible, considered as the vertical projection of $hs$, and $s''k''$ is visible as that of $sl$, whose projecting plane upon **V** also contains $hk$—$h''k''$.

Each line of the vertical projection is thus also the vertical projection of two elements.

EXAMPLE.—Assume $n$ at any point of any other element as $ef$, and $e$ at any point not on $ey$ and on any other horizontal circle except the gorge, and draw the elements containing them.

## PROBLEM XLIV.—*To represent the vertical projection of the warped surface of revolution by its meridian parallel to* **V**.

*In Space.*—1°. Pairs of elements, symmetrical with respect to any meridian plane, intersect in that meridian plane, and hence at points of the meridian in that plane (33).

2°. Points assumed on any element, and revolved around the axis of the surface, until they fall upon the same meridian plane, will also give the meridian curve in that plane. Hence—

*In Projection.*—1°. *By symmetrical elements.* Pl. X., Fig. 85. *eOy* is the horizontal trace of the meridian plane parallel to **V** and containing the required meridian curve. Then pairs of elements, symmetrical relative to this plane, meet as at $A$, $s$ and $e$, horizontal projections of points of the curve, whose vertical projections, $A'$, $s'$ and $s''$, $e'$ and $e''$, etc., are on the vertical projection, $e''s''A's'e'$, of the curve. As this curve is evidently convex towards the axis $h''h'$ of the surface, while no points of the surface in the meridian plane *eOy* can be exterior to the curve, the points $e''$, $s''$, etc., are points of contact of the vertical projections of elements which are thus *tangents to that of the meridian.*

2°. *By horizontal circles.*—Pl. X., Fig. 83. The surface being given by its axis $O—O'O''$ and the elements $ab—a'b'$ and $ab—a''b''$, assume points at pleasure as $rr'$, $ee'$, $ee''$, $bb'$, $bb''$, etc., and revolve them about $O—O'O''$, in the arcs $rn—r'n'$, $cf—e'f'$, etc., till they reach the meridian plane $mOn$, when they will be points $nn'$, $ff'$, $gg'$, etc., of the meridian curve $gfng—g''n'g'$.

Points of any other element, as $p$ on $cd$, will give other points as $qq'$ of the meridian curve.

THEOREM VII.—*The meridian curve of the warped surface of revolution is a hyperbola.*

*First.* From Pl. X., Fig. 85, the symmetry of the elements, in vertical projection, relative both to $A'B'$ and $h'h''$, and the construction of Fig. 83, alike show that the meridian curve consists of *two equal and opposite branches*, having *two rectangular axes of symmetry*, and *convex towards one of them.*

*Second.* In Fig. 85, $ab$ and $mu$, being parallel to the meridian plane, meet in it, at points of the meridian, only at an infinite distance from the gorge, but as the vertical projections of the elements are tangent to that of the meridian curve, $a'b'$ and $a''b''$ are tangent to the latter only at an infinite distance from $q'$. Likewise in Fig. 83, the further from $rr'$ the point $bb''$ is taken, the more nearly the arc $bg$

becomes a perpendicular to **V**. This occurs only when $rb$—$r'b''$ (or $r'b'$) is infinite, when $b''g'$ vanishes, and thus $r'b'$ and $r'b''$ become *tangents at infinity* to the meridian curve, that is *asymptotes to it* (51). All these properties show the meridian curve to be a hyperbola; but—

*Third.* See Fig. 84. Let $AB$—$A'B'$ be the gorge of a warped surface of revolution. Then, first, the sphere generated by the circle of the gorge is unique, being the only one having a great circle of contact with the surface; second, the asymptotes, $AB$—$a'b'$ and $AB$—$e'h$, projections of $ab$ on the meridian plane $AB$, are a unique pair of tangents, being the only ones which (Fig. 85) represent elements parallel to **V**; third, hence the tangents, as $p'F'$ and $g'f'$, at the intersections of the asymptotes with this sphere, are unique, and hence their intersections $F'$ and $f'$ with the axis $A'B'$, which is a unique line relative to the meridian curve, are a pair of *unique points* on that axis, equidistant from the centre $O'$. The hyperbola has one and only one such pair of unique points on its transverse axis, viz., its foci; moreover, drawing the semicircle on $a'F'$, we find a second right angle $F'q'a'$ having its vertex on $A'p'B'$. This agrees *first* with the property of the *hyperboloid* (Theor. VI.) that, at any point $a'$ of an element $a'O'$, one and only one other element $a'q'$ can be drawn, and, *second*, with the property of the *hyperbola* that the locus of the intersection of perpendiculars from the foci upon the tangents are on the circle described on the transverse axis. Hence we conclude that the unique points $F'$ and $f'$ are the foci of the meridian curve, which is thus further shown to be a hyperbola.

65. The warped surface of revolution having thus a hyperbola for its meridian curve, is called a *hyperboloid of revolution of two united nappes*, to distinguish it from that of *two separate nappes*, which is generated by the revolution of a hyperbola about the axis as $m'n'$ (Fig. 83) which intersects it.

The nappes are the equal and opposite halves of the

surface which meet at the gorge as the two nappes of a cone do at its vertex.

66. The hyperboloid of revolution of two united nappes, also distinguished as the warped hyperboloid of revolution, since that of two separate nappes is obviously a double-curved surface, may, from all that has now been shown, be generated in three ways: *first*, by the revolution of a straight line about an axis not in the same plane with it; *second*, by the motion of a straight line on three fixed straight lines at equal perpendicular distances in the same plane, from a fourth line; *third*, by the revolution of a hyperbola around its conjugate axis.

67. By similar reasoning to that in Theor. VII. (*Second*, Fig. 83), it may be shown that any plane parallel to the vertical planes at *dg* and *ab* and *between* them cuts the hyperboloid in a hyperbola whose transverse axis is *horizontal*, while if such a plane is in front of *ab*, that is, exterior to the gorge, it will cut the hyperboloid in a hyperbola each of whose branches is wholly on one nappe of the surface, and whose transverse axis is therefore *vertical*.

PROBLEM XLV.—*Having given either projection of a point on a warped hyperboloid of revolution, to find its other projection.*

*In Space.*—The two projections of any point of the hyperboloid are on those of any line of the surface containing the point, the element and the parallel being the most convenient ones.

*In Projection.*—Pl. X., Fig. 83, *p* being the given horizontal projection of a point, draw *pd* tangent to the gorge at *s* for the horizontal projection of the *element*, and the circle of radius *Op* for that of the *parallel* containing it. Then *p'*, now supposed to be on the upper nappe, will be found either on *d's'*, the vertical projection of the element *ds*, or on $q'p'$, that of the parallel $pp''q$. The latter is found by projecting $p''$ upon $a'b'$, or $q$ upon the meridian $d'm'q'$ and drawing $q'p'''p'$.

If $p'$ were given, $p'q'$ would be the vertical projection of the *parallel*, and a tangent $p's'd'$ from $p'$ to the meridian, and found by any known method, would be that of the *element* containing $p'p'$, whence $p$ would be on the horizontal projection $qp''p$ of the parallel, or on $dsp$ that of the element.

### B—Tangencies.

THEOREM VIII.—*Every tangent plane to a warped hyperboloid of revolution is also a secant plane, containing two elements, and is tangent at their point of intersection.*

As in (45), the tangent plane to *any* surface at any point contains all the tangent lines to the surface at that point; any two of which will determine the plane. Then (see Pl. X., Fig. 83) the tangent plane at $ff'$, for example, may be conveniently determined by the tangent to the meridian $g'v'g''$ at $ff'$, and the tangent to the circular parallel $ef$—$e'f'$ at $ff'$. But the former curve is convex towards the axis $O$—$O'O''$, and the latter is concave towards it, hence the tangent to the first is on the left, or inner side of the surface at $ff'$, and the tangent to the second is on the right, or outer side of the surface at $ff'$. These tangents being thus on opposite sides of the surface, the tangent plane determined by them is necessarily also *a secant plane*, and evidently has only a *point* of contact with the surface.

But there must somewhere be limits between those tangents at $ff'$ which, like $ft$—$f'$, are exterior to the surface, and those which, like $f'k'$, are tangent to its interior side.

These limits must necessarily lie in the surface, and being straight, must be elements of the surface. The two elements at any point (Theor. VI.) are thus merely that pair of tangents which coincide with the surface, which agrees with (44). Hence the two elements, one of each generation, at any point, considered as a pair of tangents at that point, determine most simply the tangent plane at that point.

PROBLEM XLVI.—*To construct the tangent plane at a given point of the warped hyperboloid of revolution.*

*In Space.*—By Theor. VIII., the required tangent plane will be determined by the two elements through the given point, and hence (27, 4°) its traces will be determined by their traces.

*In Projection.*—Pl. X., Fig. 86. Let $tt'$ be the given point. Then tangents to the gorge, $tp$ and $tq$, will be the horizontal projections of the determining elements, from which their vertical projections, $t'p'$ and $t'q'$, are found, and thence their traces, as $r$ and $u$ on **H**, and $P''$ on **V**. Then $PruQ$ is the horizontal, and $QP''$ the vertical trace of the required plane.

The tangent plane at $tt'$ is seen to be also a secant plane, since its horizontal trace cuts that of the hyperboloid at $r$ and $u$.

It is also perpendicular to the meridian plane $Ot$, whose horizontal trace $Ot$ is perpendicular to $PQ$. This is evident from the symmetry of the elements $tp$ and $tq$ relative to the vertical meridian plane $Ot$.

Since the tangent plane to the warped hyperboloid has only a point of contact (Theor. VIII.) while three points determine a plane, a tangent plane to it may *contain a given line*, or be *parallel to a given plane*.

EXAMPLES.—1°. Take $tt'$ anywhere on the upper nappe of the surface.

Ex. 2°. Take $tt'$ on the meridian plane $O—O'O''$.

Ex. 3°. Construct the tangent plane at a given point, by means of the tangent to the parallel at that point, and the meridian plane through the same point.

## 98   DESCRIPTIVE GEOMETRY.

**PROBLEM. XLVII.**—*To construct a plane, tangent to a warped hyperboloid of revolution, and containing a given line.*

*In Space.*—Since every tangent plane to the hyperboloid contains two elements of the surface, every other line in this plane must intersect those elements, and hence the surface, unless the line happens to contain the point of contact, when it would be a tangent. Now, let a given line $L$ intersect the surface at two points $p$ and $p_1$, and let $G$ and $g$ be the two elements (Theor. VI.) at $p$, and $G_1$ and $g_1$ those at $p_1$, $G$ and $G_1$ being of the same generation. Now, as $G$ and $g_1$ intersect both each other (63) and $L$, and as $g$ and $G_1$ do the same, two and only two tangent planes are determined by the given line $L$ and the elements at $p$ and $p_1$.

*In Projection.*—Pl. X., Fig. 86. Let $mh$—$m'h'$ be the given line, the hyperboloid being shown by its axis $O$—$O'O''$, and an element $ab$—$a'b'$, and hence also by its gorge $dpq$, and horizontal trace $abr$.

The vertical plane through $mh$—$m'h'$ cuts the hyperboloid in the hyperbola $kcd$—$k'c'd'$, found by projecting up $k$, $c$ and $d$ upon the vertical projections $a'b''$, $d'c'$ and $o'p'$ of the parallels containing these points. The given line pierces the hyperboloid at its intersection $ee'$ with this curve. Then tangents to the gorge from $e$ will be the horizontal projections of the two elements, each one of which, with $mh$—$m'h'$, will determine one of the tangent planes.

Thus the element $ef$ gives the horizontal trace $f$, and that of $mh$—$m'h'$ is $i$, hence $fi$ is the horizontal trace of one of the tangent planes. The vertical trace of this plane passes through $n$ that of $mh$—$m'h'$ and $jj'$, that of a parallel to $fQi$ at any point as $hh'$ of $mh$—$m'h'$.

A tangent to the gorge from $g$ would be the horizontal projection of the other element of the hyperboloid contained in this plane, and would intersect $ef$ in the horizontal projection of the *point of contact*. It would also intersect $mh$ in the horizontal projection of the second point in which that line cuts the hyperboloid, and thus would give no additional plane.

EXAMPLES. 1°. Find the other tangent plane containing $mh$—$m'h'$, and both projections of the point of contact of each of them.

Ex. 2°. Construct a tangent plane to a warped hyperboloid, and parallel to a given plane.

## C—Intersections.

**PROBLEM XLVIII.**—*To find the intersection of a plane and warped hyperboloid of revolution, and a tangent to the curve at a given point.*

*In Space.*—1°. *By auxiliary planes.* Intersect the given plane by any plane which will intersect the hyperboloid either in an element or a parallel; then where either of the latter intersects the line cut from the given plane by the auxiliary one, will be points of the required curve.

2°. *By auxiliary cones.*—Any line of the given plane, drawn from the intersection of the plane with the axis of the hyperboloid, will, by revolution around that axis, generate a concentric cone, cutting the hyperboloid in two parallels, and the given plane in two straight lines, elements of the cone, one of which will be the assumed one. These lines and parallels all being on the same cone must intersect. But the former being in the given plane and the latter in the hyperboloid, these intersections must be points of the required intersection of the plane and hyperboloid.

*In Projection.*—Pl. X., Fig. 87. Let the hyperboloid be given by its axis $O$—$O'O''$, and the element $ab$—$o'b'$, and let $PQP'$ be the given plane. Then—

1°. *By auxiliary planes.* Assume any horizontal plane as $R'S'$. It cuts from the plane $PQP'$ a line $qh$—$R'S'$, parallel to $QP$, and whose vertical trace is $R'$. It also cuts from the hyperboloid the parallel $R'S'$, one point of which is $s'$, on $o'b'$, which, projected upon $ab$, gives $s$, and thence the horizontal projection, with radius $Os$, of the parallel. Then the line $hq$—$R'S'$ cuts the parallel $shv$—$R'S'$ in the two points $hh'$ and $vv'$ of the required curve. Any desired number of other points may be found in the same manner.

2°. *By auxiliary cones.* We shall mention here only the one which contains the highest and lowest points. The generating line of this cone (*In Space* 2°) is that line of the given plane which is cut from it by the meridian plane perpendicular to it, since this line evidently contains those points. The cone, being tangent to the given plane, will contain no other line of it, and hence will give only the highest and lowest points, if any.

The given plane $PQP'$ cuts the axis $O$—$O'O''$ at $O,r'$ (XI. $b$, 1°), hence $Oc$—$r'c'$, where $Oc$ is perpendicular to $PQ$, is the generatrix of the cone, and the circle of radius $Oc$ is the cone's base. One point of each of the parallels in which this cone cuts the hyperboloid, if at all, is where it cuts the given element $ab$—$o'b'$.

To find the latter points, pass a plane through $ab$—$o'b'$ and the vertex $O,r'$ of the cone, and it will cut from the cone those elements which will meet $ab$—$o'b'$ in a point of each parallel.

This plane is determined (Prob. IX., 2°) by $ab$—$o'b'$ and by a parallel to it, $Od$—$r'd'$ from the cone's vertex. Hence $db$ is its horizontal trace, but it does not cut circle $Oc$, the base of the cone, and hence contains no elements of the cone; which shows that in the figure this cone does not intersect the hyperboloid. Had it done so, the construction would have been completed as follows.

1st. Note the intersection of the trace $db$ with the cone-base of radius $Oc$.

2d. Join these points with $O$, and note where they (produced if necessary) meet $ab$.

3d. Describe arcs, with $O$ as a centre, through the latter points, and their intersections with $Oc$ will be the highest and lowest points (attending carefully to the positions in space so as to note the proper intersections, which will be indicated by their vertical projections, or by other points of the curve).

In the figure, a hyperbolic intersection is shown for variety, an elliptical one being usually represented.

The transverse axis of the horizontal projection will bisect perpendicularly any chord of the curve drawn parallel to the infinite or conjugate axis $Oc$. By an analogy which holds good, between the hyperbolas in Figs. 83 and 87, as the asymptotes in Fig. 83 are parallel to the elements contained in a tangent plane parallel to the plane of the hyperbola, the like is true in Fig. 87.

3°. *The tangent line.*—Let $vv'$ be the point of contact. Then, by (Prob. XLVI.), draw the tangent plane at $vv'$, and its intersection with $PQP'$ will, by the principle of (Prob. XXXIX.), which is general, be the tangent.

EXAMPLES.—1°. Representing the hyperboloid as in Pl. X., Fig. 85, find its intersection with any plane, by constructing by (Prob. XI.) the intersections of its elements with that plane.

Ex. 2°. In Fig. 87, assume $PQP'$ in such a position as to give an elliptical intersection, and make the construction indicated above for finding its highest and lowest points.

Ex. 3°. Make the intersection, as $Oz$—$r'z'$, of any other meridian plane, $Oz$, with $PQP'$, the generatrix of a concentric cone, cutting the hyperboloid in two parallels which will meet the two elements (of which $Oz$—$r'z'$ is one) cut by $PQP'$ from this cone in four points of the curve.

68. If the hyperbola $g''n'g'$, Pl. X., Fig. 83, and its asymptotes, $a'b'$ and $a''b''$, be simultaneously revolved about the axis $O$—$O'O''$, the former will generate a hyperboloid, and the latter a cone whose vertex is $O,r'$, and which will be tangent to the hyperboloid on the circles at an infinite distance from the gorge $m'n'$, and generated by the points of contact of the asymptotes with the hyperbola. This cone is therefore called the conic asymptote, or asymptote cone of the hyperboloid.

69. The radius of the gorge of a warped hyperboloid and the inclination of its elements to the plane of the gorge are both arbitrary. If the former be reduced to zero, the

hyperboloid becomes a cone; if the latter be made 90°, the hyperboloid becomes a cylinder. The cone and cylinder are thus only particular cases of the hyperboloid.

Again, the common meridian plane of the hyperboloid and its asymptote cone (67) intersects the former in a hyperbola and the latter in a pair of intersecting straight lines, which is a particular case of the hyperbola. Any plane parallel to this meridian plane intersects both bodies in hyperbolas. The tangent plane to the asymptote cone (Fig. 83) contains the pair of parallel elements of the hyperboloid, of which $ab$—$a'b'$ is one, but a straight line, and a pair of parallels are each particular cases of the parabola. Also planes parallel to the gorge cut the two bodies in circles which are particular cases of the ellipse. All these facts lead us to infer the probability of what is really true,* that *every plane section of a warped hyperboloid of revolution is a conic section of the same kind as that cut from the asymptote cone by the same plane.*

70. Since all ruled surfaces are composed of rectilinear elements, the general problem of the intersection of the warped hyperboloid of revolution with cylinders and cones, or with other like hyperboloids, reduces to finding the intersection of a straight line with the hyperboloid;† but many special cases, as in the following examples, can be easily solved.

EXAMPLES.—1°. Find the intersection of a cylinder or of a cone with a hyperboloid when their axes are parallel.

Ex. 2°. Find the intersection of a cone and hyperboloid when the vertex of the former is in the axis of the latter.

Ex. 3° Find the intersection of a cylinder and hyperboloid when their axes are not in the same plane, while the *directions* of these axes are perpendicular to each other.

71. Even the general case, the axes of the two given surfaces having any position not in the same plane, may be

---

* See my larger Descriptive Geometry, Theorem XLII.
† See my larger Descriptive Geometry, Problem C.

solved tentatively by assuming elements of either generation of the hyperboloid, which it is supposed will contain points of the intersection, and by then passing auxiliary planes through these elements and the vertex of the cone (Prob. IX., 2°), or parallel to the axis of the cylinder (Prob. IX., 4°).

Elements, one on each given surface, and thus in the same plane, will meet in points of the required intersection of the hyperboloid with the cylinder or cone.

EXAMPLE.—Find the intersection, by the last article, of a warped hyperboloid of revolution, with either a cylinder or a cone whose axis has any position not in the same plane with that of the hyperboloid.

## D—Development.

EXAMPLE.—*To find the true size of the intersection of a plane and a warped hyperboloid of revolution, and the corresponding position of a given tangent to the curve.* [See Prob. XL.]

# PART II.—DOUBLE-CURVED SURFACES.

72. DOUBLE-CURVED SURFACES of revolution (24) are those which can be generated only by the revolution of a curved line of some kind, about an axis, and hence on which no straight line can be drawn.

The generatrix may always be a plane curve, for suppose at first that the surface has been generated by a curve of double curvature, its meridian section would necessarily generate the same surface.

73. The most important distinctions among double-curved surfaces, relative to the purposes of Descriptive Geometry, are—

*First.* Their generation by circles, or by other curves: a distinction of importance, owing to the ease of drawing tangents perpendicular to a radius.

*Second.* That of being doubly convex, or concavo-convex.

A sphere is an example of a doubly convex surface, any two plane sections whatever taken at right angles to each other through the centre being both convex towards surrounding space.

A bell is an example of a concavo-convex surface, its horizontal sections, as it hangs at rest, being convex, while its meridian sections are concave towards surrounding space.

**74.** The kinds of double-curved surfaces of revolution indicated will be sufficiently illustrated by such as can be generated by the revolution of some conic section about its own axis, or an axis in its own plane.

These surfaces are, the *sphere*, the *oblong ellipsoid* (prolate spheroid), generated by the revolution of an ellipse about its transverse axis; the *flattened ellipsoid* (oblate spheroid), generated by the revolution of an ellipse about its conjugate axis; the *paraboloid*, generated by the revolution of a parabola about its axis; and the *hyperboloid of separate nappes*, generated by the revolution of a hyperbola about its transverse axis. Also the *annular torus*, generated by the revolution of a circle about an axis exterior to it but in its own plane (73).

Two important general properties of the surfaces generated by the revolution of a conic section about either of its axes are, that all their plane sections, and their curves of contact with circumscribed tangent cylinders or cones, are conic sections.

### A—Projections.

**75.** In representing singly any double-curved surface of revolution, it is usual to assume its axis of revolution as vertical. Its *parallels* will then be in horizontal, and its *meridians* in vertical planes (33).

When such a surface is one of a group of objects variously placed, it may be necessary to represent it with its axis oblique to one or both of the planes of projection.

PROBLEM XLIX.—*To represent the projections of any double-curved surface of revolution.*

*In Space.*—The required projections will consist of those of the apparent contours, that is, of the curves of contact of two circumscribing tangent cylinders, the elements of one of which are perpendicular to **H**, while those of the other are perpendicular to **V**. Or, if, as in the case of the paraboloid and the hyperboloid, the surface is unlimited in the direction of its axis of revolution, its projection on **H**, the axis being vertical, will be indicated simply by any one of its parallels.

*In Projection.*—Pl. XI. In Figs. 90 and 91, the circles with $O$ and $O'$ as a centre represent a sphere, whose projections will always be equal circles.

Fig. 92 represents an oblong ellipsoid, its horizontal projection consisting of its greatest parallel, and its vertical one of that meridian which is parallel to **V**.

In Fig. 88 is represented an annular torus, generated by the circle, of centre $ff'$ and radius $fh$—$f'h'$, in revolving about the axis $O$—$O'O''$. The portion generated by the semicircle $m'h'n'$ is doubly convex. That generated by $m'q'n'$ is concavo-convex. The circle of radius $Oq$—$O'q'$ is the gorge.

The generating circle may be tangent to the axis, thus giving the limit at which the torus ceases to be annular; or it may intersect it.

Fig. 89 represents an oblong ellipsoid with its axis first oblique to **H** only, and then to both **H** and **V**.

We construct first the ellipse on the given axes $A'B'$ and $C'D'$. Then the horizontal projection is the base of a vertical circumscribed tangent cylinder shown by the vertical tangents at $c'$ and $d'$. The curve of contact of this cylinder being an ellipse, $c'O'd'$ is its vertical projection; hence $OE = OF = O'A'$, and the horizontal projection is the ellipse on the axes $EF$ and $cd$.

Next, suppose the ellipsoid to be revolved about a vertical axis through its centre till the vertical plane on $cd$, as shown at $c''d''$, makes any angle as $nO''c''$ with its former position. Then, first, tangents perpendicular to **V**, as at $e''$ and $f''$, will represent the cylinder which projects the ellipsoid upon **V**; and, second, no point of the surface having changed its height, the new vertical projection will still be included between the horizontal tangent planes $a'a'''$ and

$b'b'''$. Hence, projecting $a'$ at $a$, make $O''a'' = Oa$ and project $a''$ at $a'''$ and find $b'''$ likewise. Also construct $cf$ by making $cc = c''c''$, project $e$ and $f$ to $e'$ and $f'$, and thence over to $e'''$ and $f'''$, the vertical projections of $c''$ and $f''$. The ellipse $a'''f'''b'''e'''$ can be sketched with tolerable accuracy through these points of contact with its circumscribed tangents, or additional points can be found by drawing tangents perpendicular to **V**, to other sections parallel and similar to $c'd'$.

PROBLEM L.—*Having one projection of a point upon a double-curved surface, to find its other projection.*

*In Space.*—Every plane section of a sphere being a circle, assume any circular section containing the point and perpendicular to **H** or **V**, so that by revolution it may be made parallel to **V** or **H**, respectively; preferring for convenience a great circle, since this may by revolution be made to coincide with a projection of the sphere.

Or choose a section parallel to **H** or **V**.

In case of other surfaces, take either the *meridian* or the *parallel* through the given point.

In all cases, the other projection of the point will be on the other projection of the section containing it.

*In Projection.*—Pl. XI. 1°. *The point on a sphere.* Fig. 91. Let $p$ be the given projection, then $pr$ is the horizontal projection of an arc through $p$ and parallel to **V**. Supposing $p$ to be on the upper half of the sphere, $r'p'$ is the vertical projection of this arc, and $p'$ is found by projecting $p$ upon it.

A similar construction in an inverse order serves to find $q'$ when $q$ is given.

2°. *The point on any non-spherical surface.*—Fig. 93. Let $T$ be the given projection of a point on the upper half of the torus, and let $OT$ be the meridian containing it. Revolve $OT$ till parallel to **V** at $OT''$, when $T''$ will be projected at $T'''$, on the revolved meridian which coincides with that of the vertical projection of the torus. In counter-revolution,

$T''T'''$ returns to $T$ by the parallel $T''T—T'''T'$, on which, therefore, $T$ is projected at $T'$.

It is now evident that the same construction lines, omitting $OT$, which is unnecessary, represent the solution by means of a parallel, instead of a meridian through the point.

EXAMPLES.—1°. In Fig. 91, assume a vertical great circle through $p$ to find $p'$.

Ex. 2°. Assume a great circle perpendicular to **V** through $p'$, to find $p$.

Ex. 3°. In Fig. 93 let $H'$ be given, and find in various ways the horizontal projections of the four points which it represents, once by revolving the plane $O—O'O''$ about its horizontal trace into **H**.

### B—Tangencies.

76. Since a double-curved surface contains no straight line, a tangent plane can have only a *point* of contact with it.

A plane being determined by three points, a tangent plane to a double-curved surface will be determined by any three points, so conditioned as to be in such a plane, as, for example, a *given* point of contact and a point on each of any two tangents at that point (45), or by a *required* point of contact and any two exterior points, and hence by the line containing the latter.

77. As a second exterior point may have any position in respect to one first assumed, and remaining fixed, an infinite number of tangent planes may be drawn through the former—that is, through one exterior point, and tangent to a double-curved surface. These will all evidently be tangent to a cone whose vertex is the fixed point and whose elements are the tangents from that point to the given surface.

Again: a pair of exterior points may evidently be chosen in an infinite number of ways, such that the line joining them shall be parallel to a given line. Hence an infinite number of planes may be drawn parallel to a given line. These will all be tangent to a cylinder, all of whose elements are tangent to the given surface and parallel to the given line.

78. If any two lines, one of them at least curved, are tangent to each other at some point, they will, by simultaneous revolution around any one axis, generate two surfaces of revolution having a common axis, and their point of contact, remaining such, will describe a circle of contact, common to both surfaces. That is, *if two surfaces of revolution having a common axis are tangent to each other, they will have a circle of contact whose plane is perpendicular to the common axis* (57).

When one of the two surfaces is a cylinder, it will be tangent to the double-curved surface on its greatest or least parallel. A cone may be tangent on any parallel. When the parallel reaches a vertex of the surface, it becomes a point, and the tangent cone becomes a tangent plane at that point.

79. Having a plane, tangent to a double-curved surface at a given point, any single-curved surface, developable or warped, may be placed tangent to this plane so that its element of contact with it shall pass through the given point of contact. A single-curved surface will thus have only a *point* of contact with a double-curved surface.

80. *The tangent plane to a surface of revolution is perpendicular to the meridian plane through the point of contact;* for it contains the tangent to the parallel at the point of contact, and it is sufficiently evident from Fig. 92, for example, that the tangent at $nn'$, for example, to the parallel $snr$ is horizontal, and perpendicular to the meridian plane $On$, which is vertical. See also (Prob. XLVI.)

PROBLEM LI.—*To construct a plane tangent to an annular torus at a given point of contact.*

*In Space.*—The surface being generated by a circle, the distinction in (73) may be applied, and the plane found in part by the principle that it is perpendicular to the radius of the generating circle containing the point of contact.

Also the surface being one of revolution, the plane may be found on the principle of (80) as well as on the general principle of (76).

*In Projection.*—Pl. XI., Fig. 88. 1°. *Special method applicable with a circular generatrix.*—Let $tt'$ be the given point of contact, either of its projections being found by (Prob. L.) when the other is given. $ff'$, the centre of the generating circle $qfh$—$q'm'h'$, appears at $gg'$, when revolved into the meridian plane $Ot$ of the point of contact. Then $gt$—$g't'$ is the radius of this point, and as the tangent plane is perpendicular to it, its traces will be perpendicular to the projections of the radius. The *directions* of these traces thus being known, it is only necessary to find one point of each. Revolving the plane $Ot$ into **H**, $gg'$ appears at $g''$, and $tt'$ at $t''$ (Prob. XII.), and $g''t''$ is the revolved radius of contact. Then $t''k$, perpendicular to $g''t''$, is the revolved intersection of the tangent plane with the plane $Ot$, hence $k$, its horizontal trace, is a point of the horizontal trace $RS$, perpendicular to $gt$, of the tangent plane. Its vertical trace, $r'e'$, is perpendicular to $g't'$, through $r'$, the vertical trace of the tangent $tr$—$t'r'$ to the parallel through $tt'$.

2°. *General method, applicable to any surface of revolution.*

The solution can be understood by inspection of the construction, fully given, of the plane $PQP'$, tangent at $TT'$ and determined by the tangents to the parallel and the meridian at that point.

If $TT'$ were on the inner or concavo-convex portion of the torus (73) the tangent plane would (Theor. VIII.) be found to be also a secant plane.

Two tangent planes, perpendicular to the axis, will each have a circle of contact with the torus.

EXAMPLES.—1°. Take the point $tt'$ on the concavo-convex portion of the torus, retaining the method there shown.

Ex. 2°. Take the point $tt'$ on the concavo-convex portion of the torus, retaining the method shown at $TT'$.

81. Solutions of problems of tangent planes, in which an intermediate auxiliary surface is employed so that the

required tangent plane is tangent both to it and to the given surface, are *indirect*. Otherwise the solution is *direct*.

Such intermediate surfaces may by the principle of (77) be circumscribed tangent surfaces, or by that of (80) may simply have a common axis with the given surface.

**PROBLEM LII.**—*To construct a tangent plane to a sphere, and through a given line, by the direct method.*

*In Space.*—The points of contact of the two planes, which can obviously be drawn, will be those of a pair of tangents, *from* the point of intersection of the given line with a plane perpendicular to it through the centre of the sphere, and *to* the great circle contained in this plane. This plane, since the directions of its traces are known (Theor. II.), will most conveniently be determined by lines through the centre of the sphere and parallel to those traces.

*In Projection.*—Pl. XI., Fig. 90. Let $OO'$ be the centre of the sphere, and $AB=A'B'$ the given line. $Oa$—$O'a'$ and $Ob$—$O'b'$, where $O'a'$ is perpendicular to $A'B'$, and $Ob$ to $AB$, are by (Theor. II.) lines of the plane perpendicular to $AB$—$A'B'$ through $OO'$. By (Prob. XI., $b$, 2°) $AB$—$A'B'$ pierces this plane, thus given, at $cc'$. Revolving the same plane about the horizontal $Ob$ as an axis, and into the horizontal plane $O'b'$, $cc'$ will fall at $C$, by making $bC$ equal to the hypothenuse $b'''c'$ constructed as shown on $b'''y, = bc$ and $c'y$. Then $Ct''$ and $CT''$ are the tangents to the great circle of the sphere contained in the perpendicular plane, after a like revolution, and $t''$ and $T''$ are therefore the revolved points of contact of the required tangent planes.

In counter-revolution, $N$, on the axis, remains fixed, and $C$ returns to $c$, giving $cNt$ for the horizontal projection of the primitive position of $Ct''$, on which $t''$ is found by counter-revolution at $t$, the intersection of $t''t$, the arc of counter-revolution, with $ct$. Then projecting $N$ at $N'$ on the plane $O'b'$, gives $c'N'$, the vertical projection of $cN$, on which $t$ is projected at $t'$.

$TT'$ may be similarly found, except that, as in the figure, $CT''$ does not always conveniently intersect the axis $Ob$. In this case, draw any transversal $m''n''$, intersecting $Ct''$ and $CT''$, and the axis $bO$ as at $q$. In counter-revolution, $m''$ returns to $m$, on $ct$; by drawing $m''m$ perpendicular to $Ob$, $q$ remains fixed, giving $mq$, to which $n''$ similarly returns at $n$, whence it is vertically projected on $m'q'$ at $n'$. Then $cn—c'n'$ is the tangent to which $T''$ returns at $T$, by counter-revolution, in the arc $T''T$, perpendicular to $bO$, and is thence projected at $T'$.

$PQP'$ is the required tangent plane at $tt'$. Its horizontal and vertical traces are respectively perpendicular to $Ot$ and $O't'$ the radii of the point of contact (Theor. II.), and contain the like traces $d$ and $B'$, of $AB—A'B'$.

EXAMPLES.—1°. Let the centre of the sphere be in **H**.
Ex. 2°. Let it be in the ground line.
Ex. 3°. Let the given line cross any other angle than the *second*.
Ex. 4°. Let it be on the right of the sphere.

**PROBLEM LIII.**—*To construct a plane tangent to a sphere and through a given line, by means of an auxiliary tangent cone.*

*In Space.*—Make any point of the given line the vertex of a cone, tangent to the sphere on a circle of contact. Tangents to this circle from the traces of the given line on the plane of the circle will (Prob. XXX.) be the traces, on that plane, of tangent planes to the cone, through the given line. But by (77) these planes will also be tangent to the sphere, and hence will be the required ones.

*In Projection.*—Pl. XI., Fig. 91. In order that one projection of the base of the auxiliary cone may be a straight line, which will evidently be most convenient, its plane must be perpendicular to a plane of projection, and the axis of the cone must consequently be parallel to the same plane. Hence let the vertex of the cone be the trace of the given line, $AB—A'B'$ on a plane parallel to **H** or to **V** through $OO'$, the centre of the given sphere. In the figure, $aa'$ is the trace of the given line on the former plane. Then tangents $ac$ and $ad$, from $a$ to the horizontal projection of the sphere, give $acd$ for the horizontal projection of the auxiliary tangent cone, and $cd$ that of its base. $CC'$ is the trace of $AB-A'B'$ on the vertical plane, $cd$, of this base. Revolving this plane about the diameter $cd$ as an axis till horizontal, the cone's base appears as the circle on $cd$, and $CC'$ at $C''$, by making $CC''$ equal to $C'd'$.

Then $C''t''$ and $C''n''$ are the revolved positions of the traces of the tangent planes on the plane, $cd$, of the cone's base. In counter-revolution, $t''$ and $n''$, the revolved points of contact, return in vertical arcs projected in $t''t$ and $n''n$, perpendicular to the axis $cd$, to $t$ and $n$, whence their vertical projections, $t'$ and $n'$, will be found on perpendiculars to $GL$ from $t$ and $n$, and at distances $t'x'$ above, and $n'k'$ below the horizontal plane $O'a'$ equal to $t''t$ and $n''n$.

Having now the points of contact $tt'$ and $nn'$ of the required planes, the traces of the latter can be found as in Probs. LI. or LII.

EXAMPLES.—1°. Let the centre of the sphere be in **H**.
Ex. 2°. Let it be in $GL$.
Ex. 3°. Let $AB-A'B'$ cross any other than the first angle.
Ex. 4°. Let $AB-A'B'$ be to the left of the sphere.

82. By (80) if two surfaces of revolution have a common axis, without being tangent to each other, and if a plane be tangent to both of them, its points of contact with both will be in the same meridian plane, and a common tangent line to the two meridian curves contained in this plane would be a line of the common tangent plane.

But the latter plane could not contain this line and another one arbitrarily chosen, since, unless the two lines happened to intersect, they would be equivalent to *four* points, while a plane is determined by *three*.

Hence *if a plane is to be drawn through a given line and tangent to a surface of revolution by means of an exterior auxiliary concentric surface, that surface must be generated by the*

114                    DESCRIPTIVE GEOMETRY.

*given line.* Then this line, with the point of contact on the *given* surface, will be equivalent to three determining points of the tangent plane. It is further evident that the given line must not intersect the given surface.

Finally, as this method requires the given surface to have only a point of contact with the tangent plane, it applies as well to the hyperboloid of revolution (Prob. XLVII.) as to a double-curved surface of revolution.

**PROBLEM LIV.**—*To construct a plane through a given line and tangent to a double-curved surface of revolution, by means of a concentric auxiliary surface.*

*In Space.*—By (82) the concentric surface must be generated by the given line, and hence will be a warped hyperboloid of revolution (Theor. VII.). The common tangent to the two meridian curves will give one point of each of the parallels, one on each surface, containing the points of contact of the tangent planes with the two surfaces. But the point of contact on the hyperboloid is also on the given line (Theor. VIII.), and that on the given surface is on the same meridian plane with the former one (80), which determines both of them.

*In Projection.*—Pl. XI., Fig. 93. Let the given surface be the annular torus, generated by the revolution of the circle $ab$—$a'b'$ about the axis $O$—$O'O''$; and let $AB$—$A'B'$ be the given line.

Revolving a few points as $NN'$, $ss'$, $BB'$, etc., of $AB$—$A'B'$ into the common meridian plane $OB$, we find the meridian curve $B's''n'b'$ of the hyperboloid, observing that as $ON$ is perpendicular to $AB$—$A'B'$, the point $NN'$ generates the gorge, and hence determines the axis $N'n'$ of the meridian hyperbola. Also $b'$ and $t'''$ are found as symmetrical with $B'$ and $s''$ relative to $N'n'$.

Next, the common tangent $t'''T'''$, being a line of the common tangent plane, determines a point of the parallel containing the point of contact on each surface. Then revolving $t'''t''$ about $O—O'O''$ to $t't$ on $AB—A'B'$ ($t$ not shown), we have the point of contact on the hyperboloid. But both points of contact are on the same meridian plane (80); hence draw $tO$, and revolve $T'''T''$ likewise upon it at $TT'$, which is the point of contact of the required tangent plane to the torus. The traces of this plane can then be found as in Prob. LI.

As four common tangents can be drawn to the hyperbola, and to the complete meridian of the torus, which embraces the two opposite circles of centres $rr'$ and $oo'$, four tangent planes can be drawn which will fulfil the required conditions.

EXAMPLES.—1°. Find any one additional tangent plane in the above problem.

Ex. 2°. Apply the method of solution to any other double-curved surface of revolution, or to the warped hyperboloid.

PROBLEM LV.—*To construct a plane through a point in space, and tangent to a double-curved surface of revolution, on a given section of the surface.*

*In Space.*—Considering only parallels and meridians, if the section is a *parallel*, the required plane will be tangent to the cone whose base is that parallel (Prob. XXX. and 77). If it is a *meridian*, then by (80) the plane will contain a perpendicular to the plane of that meridian from the given point.

*In Projection.*—Pl. XI., Fig. 92. Let the ellipsoid, whose centre is $OO'$, be the given surface, and $AA'$ the given point.

1°. *A parallel section.*—Let $snr—m'n'$ be the given parallel, then $V''m'$ is an element of the cone having this parallel for its base, and $V''A'—OA$, the line joining the vertex with the given point (Prob. XXX.) pierces the plane $n'm'$ of the base at $p'p$, whence tangents from $p$ at $s$ and $r$ (found by a circle on $pO$) are the traces of the tangent planes from $pp'$ to the cone; and $ss'$ and $rr'$ are the intersections of their elements of contact on the cone, with the circle of contact $m'n'$ of the cone and ellipsoid, and hence are the required points of contact.

The cone becomes a cylinder at the greatest parallel $D'E'$, and $Af$ and $Ad$ are the horizontal traces of vertical tangent planes to it, and hence to the ellipsoid at $ff'$ and $dd'$.

2°. *A meridian section.*—Let $Oa$ be the plane of such a section. $Aa$ is a perpendicular to it from $A$, and $a''a'''$ is the position of $a$ after revolving about $O$—$O'V'$ into the meridian plane $OA''$ parallel to **V**. Then tangents, as $a'''b'''$, to the ellipse $h'D'E'$, from $a'''$, are the revolved positions of the traces upon $Oa$ of the tangent planes required, and $b'''b''$ is the point of contact of one of them. This point, in counter-revolution, follows the parallel through it to $bb'$, the point of contact of one of the two required tangent planes which can be drawn to the meridian in $Oa$.

In like manner, $cc'$ is the contact of one of the tangent planes to the meridian whose plane contains $AA'$; and, finally, tangents $A'h'$ and $A't'$ are the vertical traces of tangent planes, perpendicular to **V**, through $AA'$, and hence give points of contact, $h'h$ and $t't$, on the meridian $OA''$ parallel to **V**.

EXAMPLES.—1°. Find the *traces* of the tangent planes whose points of contact have now been found.

Ex. 2°. Let the given parallel be below $D'E'$.

Ex. 3°. Let $AA'$ be below $D'E'$ and on the right of the ellipsoid.

*Note.*—Problems (LII.–LIV.) will enable the student to ass at once to the construction of *curves of shade* on double-curved surfaces, in *shades and shadows* (Prob. LV.) and to that of curves of apparent contour of the same surfaces in *perspective*.

### C—Intersections.

83. A given double-curved surface may be required to intersect another given double-curved surface, or any of the three kinds of ruled surface: plane, developable, or warped.

The subject of intersections is thus here made complete, and admits the following general statement of the method of solution for all cases.

Choose such auxiliary surfaces, in respect both to form and position, as will intersect each of the given surfaces in its simplest, or most easily used, sections. Then, where the sections, cut from the two given surfaces by the same auxiliary surface, intersect each other, will be points common to both given surfaces—that is, points of their intersection.

The following problems will fully illustrate this general statement.

PROBLEM LVI.—*To find the intersection of a double-curved surface of revolution with a plane.*

*In Space.*—The best auxiliary surfaces for this case are *planes*. However placed, they will cut the given plane in straight lines, and they may be so taken as to cut the double-curved surface either in parallels or meridians, which are the two simplest ways.

That is: the intersections of any parallel or meridian of the given curved surface with the line cut from the given plane by the plane of that meridian or parallel, will be points of the required intersection. The meridian plane *perpendicular to the given cutting plane is a plane of symmetry* of this plane and of the given surface, and hence of their intersection.

I. *In Projection.*—*The intersection of a flattened ellipsoid (oblate spheroid) by a plane.* Pl. XII., Fig. 95.

1°. *The method of meridians.*

The ellipsoid is represented by the projection of the circle $OA$, its greatest parallel, and of the ellipse $AB$—$A'B'C'D'$, its meridian parallel to **V**. $PQP'$ is the given plane.

$Oh$, perpendicular to the horizontal trace $PQ$, is the like trace of the meridian plane of symmetry, and it cuts $PQP'$

in the line $Oah$—$p'a'h'$, whose vertical projection is determined by $a'$, vertical projection of $a$, and $p'$ that of the point where $PQP'$ cuts the axis $O$—$C'D'$ of revolution, and found, as shown, by (Prob. XI., $b$, 1°).

Revolve the meridian plane $Oh$ about $O$—$C'D'$ till it becomes parallel to **V**. This meridian will then coincide with $OA$—$A'B'C'D'$, and the line $Oa$—$p'a'h'$ will appear at $Oh''$—$p'a'''h'''$, giving $h'''$ and $l''$ as the revolved positions of the highest and lowest of the required points. Their primitive positions, $hh'$ and $ll'$, can then be found in either of the two ways already shown for the like points in (Prob. XXXV., 2°).

This method can be applied with any other meridian plane, but requires a like revolution and counter-revolution, in every case except that of the meridian $OA$, parallel to **V**. The plane $OA$ cuts $PQP'$ in the line $Of$—$p'f'$, where $p'f'$ is parallel to $P'Q$, and gives at once $k'k$ and $g'g$ ($g$ not shown) as the points in the plane $OA$.

2°. *The method of parallels.*—This is most convenient for finding additional points as they are found without the process of revolution and counter-revolution.

Thus, assuming any horizontal auxiliary plane as $c'd'$ between $h'$ and $l'$, it cuts from the ellipsoid the parallel of radius $Ob$, $=c'b'$, and from $PQP'$ the horizontal $c'd'$—$de$ ($de$ parallel to $PQ$) which intersects the circle $Ob$—$c'b'$ at two points of the required curve, one of which is $e$, thence vertically projected at $e'$.

Other points may be similarly found. To avoid confusing the figure, the curve, one half of which passes through $hkegl$ and $h'k'e'g'l'$, is not shown.

II.—*The plane sections of the annular torus.*—Pl. XII., Fig. 94. These, all found in the way just described, are of a great variety of forms, some of the general characteristics of which may be determined in advance by observing the relation of the line $I$, intersection of the given cutting plane with the meridian plane of symmetry, to the meridian $M$, contained in the latter plane. This line $I$ may have numerous different tangent or secant positions, as *ij*, *xy*, *hl*, etc., relative to *M*, each of which will indicate a new form of the intersection, while an exterior position, as *mn*, shows that the given plane does not intersect the torus.

A plane, like $PQP'$, Pl. XII., Fig. 94, tangent at two points, $tt'$ and $TT'$, is called a *bi-tangent plane*.

THEOREM IX.—*The section of an annular torus, made by a bi-tangent plane, consists of a pair of circles whose diameter is that of the generating circle plus that of the gorge.*

Pl. XII., Fig. 94.—Draw the radius $A'E'$ of the generating circle to the extremity $E'$ of the diameter of the parallel containing any point, as $uu'$, of the curve, also the radius $O'E'$ of the torus, to the same point $E'$. Suppose the curve *pade*, found as in the last problem, to be revolved about the line $BD—P'Q$ as an axis, till parallel to **V**. Then $uu'$ will fall at $u''$ on the perpendicular $u'u''$ to $P'Q$, at a distance $u''u'$ from $P'Q$, equal to that of $u$ from $BD$. Draw $O's$ perpendicular to $E'A'$ produced; $O'x''$ perpendicular to $P'Q$; $E'F$ and $u'v$ perpendicular to $O'A'$; $T'x''$ perpendicular to $O'O''$; and finally draw $u''x''$. In the figure, $uv$ happens to be very nearly tangent to the circle $A'D'$, because $P'Q$ makes very nearly an angle of 45° with $GL$.

The triangles $A'E'F$ and $A'O's$ are similar, and so are $O'u'v$ and $O'A'T'$. These pairs give respectively the proportions,

$$A'E' : A'O' :: E'F : O's$$
and
$$A'T' : A'O' :: u'v : O'u'$$

whence, by comparing terms, we see that

$$O's = O'u'$$

also $O'u'' = O'E'$; hence the triangles $O'u'u''$ and $O'E's$ are equal, the angle $O'E'A'$ is equal to the angle $O'u''u'$, and hence to $u''O'x''$. Comparing the triangles $O'u''x''$ and $O'E'A'$, we thus find equal angles included between equal sides, since $O'x'' = A'T' = A'E'$. Hence the remaining sides $u''x''$ and $O'A'$ are equal. But $O'A'$ is constant for all positions of $uu'$, hence the different positions of $u''$, revolved positions of different points as $uu'$ on the curve, are at a constant distance from the fixed point $x''$, and thus form a *circle*, of a radius equal to $O'A'$.

There are evidently two such circles in the plane $PQP'$, and another like pair in a plane, tangent at $t''$.

The horizontal projections of the former are ellipses, in each of which the transverse axis, as $pd$, equals the sum of the diameters of the gorge and generating circle, and the conjugate axis, as $ae$, is the horizontal projection of $a'e'$. But $Oa = Oe = O'A' = O''B'$, the radius of the highest and lowest parallels. But $O'A' = \frac{1}{2} I'D' = \frac{1}{2} pd$. That is, $Oa = \frac{1}{2} pd$, hence $O$ is a focus of the ellipse $pade$, and hence also of the symmetrical one $qbc$.

The revolution of the tangent $BD—P'Q$ about $O—O'O''$ generates a bi-tangent cone, having the parallels through $T'$ and $t'$ for its circles of contact, one on each nappe. Every tangent plane to this cone cuts the torus in a pair of circles equal to those just described.

The annular torus has therefore three sets of pairs of circular sections. 1st. Its parallels. 2d. Its meridians. 3d. Those in tangent planes to the bi-tangent cone. All the circles in each of the two latter sets are equal.

EXAMPLE.—Construct a series of figures of intersections of an annular torus, with cutting planes whose intersections with the meridian plane perpendicular to it shall have every possible variety of position relative to the meridian in the latter plane. When $PQP'$, Fig. 94, revolving about $Od—O'$ becomes so nearly vertical that $T, t$ and $d$ shall all be on the same side of a tangent parallel to V at $q$, the horizontal projections of the intersections will be natural egg-formed ovals—that is, those not compounded of circular arcs.

PROBLEM LVII.—*To find the intersection of a single-curved surface with a double-curved surface, both of revolution.*

*In Space.*—This comprehensive problem embraces the three general cases in which the single-curved surface may be a *cylinder*, a *cone*, or a *warped hyperboloid;* and each will present special cases depending on the relative position of the axes of the given surfaces.

1°. For all cases whatever in which the given axes are *parallel*, points generally of the intersection will be best found by means of auxiliary planes perpendicular to the two axes, each of which will thus contain a parallel of each surface. The intersections of these parallels will be points of the required curve. The plane of the axes will also readily give certain points.

2°. The same method can also be followed in case of a cylinder whose axis is in a plane perpendicular to that of the double-curved surface.

3°. When one of the two surfaces is a cone whose axis intersects that of the double-curved surface at any angle, the auxiliary planes may be any meridian planes limited by those which are tangent to the cone. Such a plane will cut a pair of elements from the cone (one in case of the tangent plane) and a meridian from the other surface, whose intersection will be four points (two for the tangent plane) of the required curve.

4°. For all other cases in which the axes intersect, their point of intersection should be made the centre of a system of auxiliary spheres, each of which will (57) intersect each surface in a circle, and the intersection of whose circumferences will be two points of the required intersection.

5°. For all cases in which the axes are not in the same plane, the intersection of *each* auxiliary plane with one of the given surfaces must be found as in Prob. XXXVI. or XLVIII. or LVI.

EXAMPLES.—1°. Draw on a larger scale and explain the construction given in Pl. XII., Fig. 95 (the right-hand portion), the intersection of the ellipsoid $AB-A'B'C'D'$ with the vertical conic frustum whose bases are $EHF-E'F'$ and $ZK-Z'o''$.

Ex. 2°. Find the intersection of the cone and sphere shown in Pl. XII., Fig. B.

Ex. 3°. Find the intersection of the cylinder and torus in Fig. C.

Ex. 4°. Find the intersection of an oblong ellipsoid and warped hyperboloid whose axes are parallel.

Ex. 5°. Find the intersection of an annular torus and a cylinder whose axis is in a plane perpendicular to that of the torus.

## PROBLEM LVIII.—*To find the intersection of two double-curved surfaces of revolution, whose axes intersect.*

*In Space.*—As readily appears on examination, the consideration which determines the solution is, that the surfaces are of revolution, and not that they are double-curved. Hence the solution of case 4° of the last problem applies here, and serves as a general illustration for all such cases.

*In Projection.*—Pl. XII., Fig. 96. Let the given surfaces be the flattened ellipsoid (oblate spheroid) $OA$—$C'A'D'$, and the paraboloid whose vertical projection, only, $s'r'b'$, is shown, that alone being needed. **V** is taken for convenience parallel to the plane of the axes $C'D'$ and $S'B'$.

Then make the point $O,S'$, the intersection of the axes, the centre of any convenient number of auxiliary concentric spheres; $S'f'$ is the radius of one of these spheres, and the arc $f'k'$ the vertical projection of a portion of its great circle in the plane $OA$ of the given axes. It therefore (57) cuts from the ellipsoid a horizontal circle, projected in $g'h'$ and the circle of radius $Og$ equal to $h'g'$; and from the paraboloid a circle perpendicular to **V**, and vertically projected in $e'f'$. These two circles intersect each other in a common chord perpendicular to **V** at $m'$, a chord whose extremities, projected at $m'$, being the intersections of the circumferences of the circles, are points of the intersection of the given surfaces. Thus by projecting $m'$ at $m$ and $n$ on the circle $Og$, we have two points, $nm'$ and $mm'$, of the required intersection.

Other points may be similarly found. $rr'$ and $ss'$, the intersections of the meridians in the plane $OA$, are the highest and lowest points of the curve.

EXAMPLES.—1°. Find the intersection of the ellipsoid and paraboloid. Pl. XII., Fig. D.
Ex. 2°. In Ex. 1 substitute for the paraboloid another ellipsoid.
Ex. 3°. Let one of the two surfaces be an annular torus.

**PROBLEM LIX.**—*To find points of the intersection of two ellipsoids of revolution whose axes are not in the same plane.*\*

*In Space.*—Let $E$ and $E_1$ be the given ellipsoids, and take **V** for convenience parallel to the axes of both. Then make the centre of $E$, for example, the centre of an auxiliary ellipsoid $E_2$, similar to $E_1$, but with the radius of its greatest parallel equal to the like radius of $E$. Then $E_2$ will intersect $E$, if at all, in an ellipse $e$, whose plane, $P$, will cut $E_1$ in an ellipse $e_1$, similar to $e$, because $E_2$ and $E_1$ are similar. Now, make $e$ and $e_1$ the directrices of similar and parallel cylinders $C$ and $C_1$, so inclined to **H** that their horizontal traces shall be circles. The intersections of these traces will be the like traces of the parallel elements $s$ and $s_1$ in which these cylinders will intersect each other; and, finally, the intersections of these elements with the ellipses $e$ and $e_1$ will be the points common to $e$ and $e_1$,—that is, points common to $E$ and $E_1$, and thus points of their required intersection.

Other planes $P_1$, etc., parallel to $P$, will cut the given ellipsoids $E$ and $E_1$ in ellipses similar to $e$ and $e_1$, from which, by repeating the same operations as before, other points can be found.

*In Projection.*—Pl. XII., Fig. 97. The construction being fully given, can be understood from the above description.

**PROBLEM LX.**—*To construct a tangent line at a given point of the intersection of two surfaces of revolution, one or both of which is double-curved.*

*In Space.*—1°. If one of the given surfaces is a *plane*, the required tangent will be the intersection of that plane with a plane tangent to the given double-curved surface at the given point.

2°. If both surfaces are curved, the required tangent will be the intersection of two planes, each tangent to one of the surfaces at the given point.

---

\* From Géométrie Descriptive by L̲ᴇꜰᴇʙᴜʀᴇ ᴅᴇ Fᴏᴜʀᴄʏ.

When one of the surfaces is warped, or when both are double-curved, the auxiliary tangent planes to such surfaces are conveniently made tangent to an auxiliary tangent cone of revolution having the parallel containing the given point for its circle of contact with the given surfaces.

3°. When, however, the axis of one of the surfaces is not perpendicular either to **H** or **V**, the tangent plane to it is less conveniently found than in the previous problems of tangency, and the following method becomes more suitable.

*Method by Normals.*—The normal to any curved surface at a given point is there perpendicular to the tangent plane at that point. Therefore, there can be but one normal at any one point. But there can be a normal to each of any two surfaces at any point common to both, and these two lines determine what is called the common-normal plane at that point. Every tangent line to either surface at the same point being evidently perpendicular to the normal to the same surface, the required one, which is common to both tangent planes, is perpendicular to both normals, and hence to their plane.

*In Projection.*—1°. *A tangent to the intersection of a plane and annular torus.*—Pl. XII., Fig. 94. Let the tangent be constructed at the point $f, T'$. The line $r'R'$ being tangent at $r'r$ to the meridian $T'r'$, generates the auxiliary tangent cone on the parallel containing the point $fT'$. Then the circle of radius $OR$ is the horizontal trace of this cone, and $IY$, tangent at $I$, the foot of the element through $f$, is the horizontal trace of the auxiliary tangent plane to this cone. $Y$, the intersection of $IY$ with $YQ$ the horizontal trace of the given plane $PQP'$, is a point of the tangent (1°), hence $Yf—QT'$ is the required tangent. At $TT'$ the two auxiliary tangent planes coincide, and the solution fails; but, knowing that the curve *pade* is an ellipse, a tangent at any point of it can be readily found.

2°. *A tangent to the intersection of the ellipsoid and paraboloid.*—Pl. XII., Fig. 96. Let the tangent be drawn at $ii'$. Then, adopting the *method by normals*, $i''$ and $I$ are the positions of $ii'$, after revolution about the axis of each body into the plane $OA$, and $i''N'$ and $Io'$, perpendicular to tangents at $i''$ and $I$, are the revolved positions of normals, one to each body at $ii'$. Then $N'o'$ is the trace of the normal plane at $ii'$ on the plane $OA$; hence, as this plane is parallel to **V**, the line $t'i'$ (only partly shown) perpendicular to $N'o'$, is the vertical projection of the required tangent. By counter-revolution, $Oi$—$N'i'$ and $oi$—$o'i'$ become the projections of the real positions of the normals. Hence, projecting $v'$ and $w'$, where $N'i'$ and $i'o'$ meet $O'A'$, upon $Oi$ and $oi$ at $v$ and $w$, we find $vw$ the trace of the normal plane on the plane $O'A'$, for example, chosen simply as a plane parallel to **H**. Hence $tiy$ perpendicular to $wv$, is the horizontal projection of the tangent.

EXAMPLES.—1°. Find the tangent at any point of the intersection of the plane and ellipsoid. Pl. XII., Fig. 95, or in Fig. A.

Ex. 2°. Construct the tangent at a given point of each of the intersections required in Figs. B, C, and D.

Ex. 3°. Draw on a larger scale, and explain the given construction of the tangent $Su$—$S'u'$ at a given point $uu'$ of the intersection of the ellipsoid and frustum. Pl. XII., Fig. 95.

# BOOK II.—SURFACES OF TRANSPOSITION.

## PART I.—RULED SURFACES.

### CHAPTER I.

#### PLANE SURFACES.

84. WHEN the moving line which is the generatrix of any surface moves in any other way than by revolution about a fixed axis, its motion is distinguished as one of *transposition*, and the resulting surface as a *surface of transposition* (20).

85. We have already seen (22) that *a plane is truly a surface of revolution;* but a plane can also be generated by the motion of a straight line in any manner, or not parallel to itself, upon two fixed straight lines which intersect. Such motion being evidently one of transposition, *a plane may also be regarded as a surface of transposition.* Accordingly, we shall here present as examples a few principles and problems in which planes are considered from this point of view.

#### A—Projections.

PROBLEM LXI.—*To draw lines through a given point which shall determine a plane parallel to a plane which is given by means of two lines in it.*

*In Space.*—The required lines will simply be parallel to any lines, each of which intersects both of the two given lines.

*In Projection.*—Pl. IV., Fig. 31. Let $ab$—$a'b'$ and $cd$—$c'd'$ be the lines which determine the given plane, and let $xx'$ be the point through which a parallel plane is to be passed. Draw any lines whatever, as $ed$—$e'd'$ and $en$—$e'n'$, not necessarily through the same point $ee'$, but only so as to intersect both of the given lines (85). Then at $xx'$ draw lines $xd_1$—$x'd_1'$ and $xn_1$—$x'n_1'$ parallel to those thus drawn in the plane given by $ab$—$a'b'$ and $cd$—$c'd'$, and they will be the required determining lines of a plane through $xx'$ and parallel to the given plane.

EXAMPLES.—1°. Let the point $bb'$ be in any other angle than the first.
Ex. 2°. Let $xx'$ and $bb'$ be in different angles.

### B—Tangencies.

PROBLEM LXII.—*To find the traces of a plane which is given by two lines, neither of which intersects the planes of projection within convenient limits.*

*In Space.*—Let $A$ and $B$ be the two lines. Take any point $M$ on $A$ and any point $N$ on $B$, then the projections of the line $MN$ joining these points will be $mn$—$m'n'$, and the traces of this line will be points of the respective traces of the plane. Two such lines, each intersecting both of the given lines, will determine the required plane, and evidently in a way agreeing with the definition of the plane as a *surface of transposition* (85). See also (27, 9°).

*In Projection.*—EXAMPLES.—1°. Solve the problem when the plane is regarded as a surface of transposition.
Ex. 2°. Solve the problem when it is regarded as a cylinder.
Ex. 3°. Solve the problem when it is regarded as a cone.

### C—Intersections.

86. Lines intersecting given lines which lie on the same plane are called *transversals* relative to the latter. The following principles and problems will partly illustrate the connection of descriptive geometry—that is, the method of projections—with the subject of transversals.

**PROBLEM LXIII.**—*To find the intersection of two planes, each of which is given by its horizontal trace and one point.*

*In Space.*—Any plane containing the given points and intersecting the given traces will cut each plane in a line, joining the given point of that plane with the point cut from its trace. The intersection of these lines will be a point of the intersection of the planes.

*In Projection.*—Pl. XIII., Fig. 98. Using the peculiar notation of Olivier* once for illustration: Let $H^P$ and $H^Q$ be the given traces of the planes $P$ and $Q$. Let $a^h$, $a^v$ be the projections of the given point $a$ of the plane $P$, and $b^h$, $b^v$ those of $b$, that of the plane $Q$.

---

* Designed to be completely systematic and self-explaining, to facilitate thus the reading of projections.

I.—*On the primitive* **H** *and* **V**.

$a^h$ = horizontal projection of point $a$.
$a^v$ = vertical  "    "    "    "
$A^h$ = horizontal  "    "    line $A$.
$A^v$ = vertical  "    "    "    "
$H^P$ = horizontal trace of plane $P$.
$V^P$ = vertical  "    "    "    "

Common System.

$aa'$ = projections of a point $A$.
$ab$—$a'b'$ = projections of a line $AB$.
$PQP'$ = traces of a plane.
$a''$, $a'''$, etc., or $a_1$ $a_1'$, etc., = the projections of a point $A$ either on *new planes* or after a *rotation*.
$P_1Q_1P_1'$, or $P''Q'P'''$, etc., = the like for planes.

II.—*On new planes of projection.*

$a^{h'}$ = projection of the point $a$ ($a^h$ $a^v$) on a plane perpendicular to **V** and taken as a new plane **H**.
$a^{v'}$ = projection of the point $a$ ($a^h$ $a^v$) on a new vertical plane.
$A^{h'}$, $A^{v'}$ = projections of a line $A$ on a new **H** or **V**, respectively.
$H'^P$ = the trace of the plane $P$ on a new plane **H** perpendicular to **V**.
$V'^P$ = the trace of the plane $P$ on a new plane **V**.

III.—*After rotation about an axis.*

$a'^h$, $a'^v$ = the projections of the same point $a$ as before, after rotation about an axis perpendicular to either **H** or **V**.
$A'^h$, $A'^v$ = the projections of the line $A$, after a like rotation.
$H'^P$, $V'^P$ = the traces of the plane $P$, after a like rotation.

Various causes, among them the large number of slightly different symbols, the difficulty of using them on crowded diagrams, and the necessary recourse to numbers instead of letters in large and complicated figures, have prevented this notation from coming into general use.

Draw the line $D$—that is, the line whose projections are $D^h$ $D^v$—and find its horizontal trace $n^h$. Then any line, $H^x$, through $n^h$, and intersecting $H^p$ and $H^q$, will be the horizontal trace of a suitable auxiliary plane $X$. The plane $X$ cuts from the plane $P$ the line $B$ ($B^h B^v$), and from $Q$ the line $C$. These lines intersect at $p$, a point of the required intersection $I$; also $m$, the intersection of the given traces, is another point of $I$.*

87. We know that the intersection of two planes is a straight line passing through the intersections of their like traces, and that the last problem gives a correct construction of that line. That is, regarding now the horizontal projection alone, the point $p^h$ will always be on the line $I^h$ drawn through $m^h$, however the line $H^x$ is drawn. That is, we have the following:

THEOREM I.—*Of transversals.*—If any three lines, as $D^h$, $H^r$, $H^q$, cut each other two and two, and if we have three points, $n^h$, $a^h$, $d^h$ on one of them, $D^h$, and then draw from one of these points, as $n^h$, any number of transversals, as $H^x$, cutting the lines $H^p$ and $H^q$, in points which shall be joined with $a^h$ and $b^h$ respectively by straight lines, the intersections of these last lines with each other will, together with $m^h$, fall on a straight line $I^h$.

In like manner, $D^h$, $H^p$, $I^h$ might have been the given lines, and $a^h$ the origin of the transversals, and a like series of final intersections would have fallen on a line $H^q$ containing $m^h$. Or $D^h$, $H^q$, $I^h$ being given, and $b^h$ taken as the origin of transversals, a like series of intersections would have determined the line $H^p$.

PROBLEM LXIV.—*To find the intersection of two planes whose traces do not meet within the limits of the figure.*

*In Space.*—Intersect the given planes by auxiliary planes

---

* Observe that in this notation, a point, line, or plane is *named by the single letter which indicates it in space*, and which with the indices $h$ and $v$ indicates the *projections* of the two former, or with $H$ and $V$, indicates the *traces* of the latter.

parallel to the ground line, then the pair of lines cut from the given planes by any one of the auxiliary planes will meet at a point of the required intersection.

*In Projection.*—Pl. XIII., Fig. 99. Returning to the usual notation, let $PQP'$ and $RSR'$ be the two given planes, and let $TU$ and $P'R'$ be the traces of an auxiliary plane taken nearly parallel to **V** as well as parallel to $GL$. Then $Tp$ and $Ur$, the horizontal projections of its intersections with the given planes, meet at $n$, a point of the horizontal projection of the given planes. Another such point, $m$, is found by means of a similar plane, $VW, P'R'$; then $mn$ is the horizontal projection of the required intersection. Its vertical projection could be similarly found by similar auxiliary planes making a small angle with **H**.

88. As in Prob. LXIII., we know that the last construction correctly gives the intersection of two planes, and that this line is straight and passes through the intersection of $QP$ and $SR$. Examining then the horizontal projection only of Fig. 99, as a plane figure, we deduce the following:

THEOREM II.—*Of transversals.*—If we have three lines, as $PQ$, $QS$, $SR$, which intersect in three points, and two points, as $p$ and $r$, on one of them, $QS$; and if we draw a series of transversals, as $TU$ and $VW$, parallel to $QS$, and draw lines from their intersections with $PQ$ to $p$, and from those with $RS$ to $r$, the pairs of lines, as $Tp$ and $Ur$, thus found, will meet on a straight line $mn$ containing the intersection of $PQ$ and $RS$.

This theorem is in fact that case of the last one (87), in which $n^h$, Fig. 98, is at infinity on the line $D^h$, in which case the different positions of $H^x$ would be parallel to $D^h$.

Returning to the figure in space, this case is that in which the points $a$ and $b$ being at equal heights above **H**, the line $D$ ($D^h$, $D^v$) would be horizontal, which would carry $n^h$ to infinity, and would make the solution the same as that of Fig. 99, except that each auxiliary plane, instead of being given by its traces, would be given by two lines, one of which would be its horizontal trace, while the other would contain the two given points of equal altitude.

89. From Theorems (I. and II.) the following reciprocal one is now derived:

THEOREM III.—*Of transversals.*—If there be four straight lines, $D$, $P$, $Q$, $I$, Fig. 98, three of which meet at a point, and each cuts a fourth one, and if we join any number of points on one of the former, as $I$, with two points, as $a$ and $b$, on the fourth, $D$, these lines from $a$ will cut $P$, and those from $b$ will cut $Q$, so that lines joining the pairs of points, as $c$ and $d$, thus found on $P$ and $Q$, will all meet at one point of $D$, as $n'$, Fig. 98, or be parallel to it, as in Fig. 99.

90. If, in Pl. XIII., Fig. 99, the auxiliary planes had been parallel to each other, $Vp$ would have been parallel to $Tp$, and $Wr$ to $Ur$, but $m$ would evidently still have remained on the line $mn$. That is:

THEOREM IV.—*Of transversals.*—If two given lines, as $PQ$ and $RS$, Fig. 99, be cut by a series of parallel transversals, as $QS$, $TU$, $WV$, and if through the intersections of these parallels with the given lines we draw two series of transversals, as $Tp$ and parallels to it from $Q$ and $V$; and $Ur$ and parallels to it from $S$ and $W$; the pairs of lines, as $Tp$ and $Ur$, thus drawn will intersect on a straight line, as $mn$, which contains the intersection of the given lines $QP$ and $SR$.

PROBLEM LXV.—*Having given two lines and a point, to construct a third line through this point so that the three lines shall all meet in one point.*

Pl. XIII., Fig. 100. Let $AC$ and $EF$ be the given lines, and $m$ the given point. Draw any line, $AD$, and then $Amb$ and $amB$. Draw $Bb$ and note its intersection $D$ with $AD$. Then draw $DC$, $cB$, and $Cb$; the intersection of the last two will give $n$, which, with $m$, will determine the required line.

Passing from this plane construction to its original in space, $AC$ and $EF$ may be the horizontal traces of two planes whose intersection is required, and which are given, as in Prob. LXII., by these traces and the points $B$ of the plane $E$ and $b$ of the plane $A$. Then $D$ being the horizontal trace of the line $Bb$ in space, any transversals, as $DA$ and $DC$, will represent the horizontal traces of auxiliary planes, the former cutting from the two planes the lines $Ab$ and $aB$, whose intersection is one point of that of the given planes, and the latter, $DC$, cutting them in the lines $Cb$ and $cB$, which give another point, $n$, of the intersection $mn$, which being thus evidently the required intersection, necessarily contains the point of concurrence of $AC$ and $EF$.

91. The last problem permits the enunciation of

THEOREM V.—*Of transversals.*—Having any two lines, $AC$ and $EF$, and a point, $D$, in their plane, if any number of transversals be drawn from $D$ to the given lines, also diagonals of the quadrilaterals, as $ABab$, thus formed, the intersections, as $m$ and $n$, of these diagonals and that of the given lines will all be on the same line $L$.

These theorems of transversals show how plane and solid geometry mutually assist each other, being derived from constructions in space, while in the explanation of the last problem we return from a plane construction to one in space.

PROBLEM LXVI.—*Through a given point, to draw a line that shall intersect two given lines.*

*First solution—In Space.*—Find where the plane containing the point and one of the given lines cuts the other given line. The line through this and the given point will be in this plane, and will therefore evidently fulfil the required conditions.

*In Projection.*—Pl. XIII., Fig. 102. Let $pp'$ be the given point, and $AB$—$A'B'$ and $CD$—$C'D'$ the given lines. The auxiliary plane contains $pp'$ and $CD$—$C'D'$, as is shown by its being determined by the lines $pC$—$p'C'$ and $pb$—$p'b'$, both of which intersect $CD$—$C'D'$. This plane intersects

$AB-A'B'$ at $mm'$ (Prob. XI., $b$, 2°), hence $mpn-m'p'n'$ is the required line.

*Second solution—In Space.*—This problem can also be solved by two auxiliary planes, each of which contains the given point and one of the given lines. The intersection of these two planes will be the required line, and the construction is evidently the same as that for finding where a given line pierces a plane which is given by a point and a line.

*In Projection.*—Pl. XIII., Fig. 101. Considering the construction from the point of view last named, we will examine the peculiar case in which the two traces of the given plane coincide, as at $PQP'$, and in which also the two projections of the given line coincide in the line $OB$. Then applying (Prob. XI., $a$) this line pierces that plane at a point whose two projections coincide in $m$.

A line whose projections thus coincide is in the bisecting plane of the *second* and *fourth* angles, hence $m$ is also in that plane. Therefore if a series of such lines as those whose projections coincide in $OC$, $OD$, etc., be passed through $O$ where the given line $OB$ meets the ground line, they, and consequently their intersections with $PQP'$, all found by Prob. XI., will be in the same bisecting plane. But these points are also in the plane $PQP'$, hence in the intersection of these two planes. That is, they are all in the same straight line.

92. The last problem, looking upon its construction, Fig. 101, as a plane figure, gives the following:

THEOREM VI.—*Of transversals.*—If two lines, $GA$ and $DA$, intersect at a point $A$, and if from a point $O$ on $GA$ a series of lines, $OB$, $OC$, $OD$, be drawn, cutting the other line, $DA$, as at $B$, $C$, $D$, and if from the latter points perpendiculars, $BE$, $CF$, etc., be drawn to $GA$, and if $Oo$ be drawn, also perpendicular to $GA$, the lines $oE$, $oF$, $oG$ will cut $OB$, $OC$, $OD$ in points $m$, $n$, $p$, which, with $A$, are on the same straight line.

The perpendiculars to $GA$ at $E$, $F$, $O$, $G$ may have any other direction if only they remain parallel.

## CHAPTER II.

### DEVELOPABLE SURFACES.

*Definitions, etc.*

93. THERE are two kinds of developable surfaces:
1°. Those in which *any* two elements are in the same plane.
2°. Those in which only consecutive ones are so situated.

Cylinders and cones form the first of these two kinds of developable surfaces.

94. *A cone of transposition*—that is, a cone defined in the most general manner—is the surface formed by joining a fixed point with all the points of any fixed curve of any kind whatever, and is therefore generated by a line, moving, so as always to contain a fixed point, the vertex, and upon a fixed curved directrix. *Any* two elements are therefore in the same plane. When the vertex is at infinity, the elements become parallel and the surface becomes a cylinder.

When the linear directrix is straight, the cone of transposition becomes a plane by the above definition, as the cone of revolution does by (22). See Prob. LXIII., where the intersection of the given planes is really found exactly as in Prob. XXXVII.

95. The oblique cone, and cylinder, with a circular base, are types of the entire class of cones and cylinders of transposition (20) just defined.

Pl. XIII., Fig. 103, shows, at $VAB—V'C'D'$, an oblique cone with a circular base, more briefly termed a *scalene cone*, in contrast with the cone of revolution $Vgd—V'r'ni'$. Its section perpendicular to the axis is an ellipse.

96. *The sub-contrary section.*—Pl. XIII., Fig. 103. $VO$, the line from the vertex to the centre of the circular base of a scalene cone, is its axis. A plane perpendicular the base and containing the axis is called a *principal plane*, and contains evidently the longest and shortest elements, or those respectively of least and greatest inclination to **H**.

Neglecting now the horizontal projection, suppose $V'C'D'$ to be a scalene cone with circular base whose principal plane is parallel to **V**. Then $Xx$ parallel to **H** will be a circular section. Now let $Yy$ be a section such that the angles at $Y$ and $x$, and hence all the angles of the triangles $XYz$, and $xyz$, are equal. Then

$$Xz : yz :: Yz : xz \quad \text{hence}$$
$$Xz \times xz = Yz \times yz.$$

But the section $Xx$ is a circle, hence $Xz \times xz$ equals the square of the ordinate at $z$; but this ordinate is common to the two sections, and hence also equals $Yz \times yz$. Hence the section $Yy$ is also a circle, and relative to $Xx$ is called a *sub-contrary section*.

97. *The second kind of developable surfaces* embraces innumerable surfaces formed as follows:

In Pl. XIII., Fig. 104, let $bd$ represent a line intersecting $eg$ at $d$, not in the plane $PQ$, but in space. Likewise let $eg$ intersect $hj$ at $g$, let $hj$ intersect $kl$ at $j$, etc. Also let $acfi$.... be a curve tangent to these lines as at $c$, $f$, $i$, etc. This curve will be of double curvature (49), since the lines $bd$, $eg$, etc., are not all in one plane.

Now, when these lines become consecutive, the intersections $d$, $g$, etc., will be confounded with the points of contact $c$, $f$, etc., and the lines will constitute a surface whose elements are all tangent at consecutive points of a curve of double curvature.

**98.** *This surface is developable since its consecutive elements intersect*, though not all at the same point as in a cone. This may be more explicitly shown as follows: *bd* may be revolved about *eg* as an axis into the plane *egh*, this plane may then be revolved about *hj* as an axis till it coincides with the plane *hjk*, thus bringing the lines *bd*, *eg*, *hj*, *kl*, all into the same plane. This process, equally applicable when these lines become consecutive, being continued indefinitely, all the elements will be brought into one plane without changing the relative position of consecutive ones. That is, the surface is developable.

**99.** There being an infinite number of different curves of double curvature (50), there will be an infinite variety of developable surfaces of the kind just described.

The directrix *acfi* . . . being a line, from every point of which two elements diverge in opposite directions, it is analogous to the vertex of a cone, *first* in being the limit between the *lower nappe, acfiabeh,* and the upper one *acfipqr;* and, *second*, in being the place where the surface is most contracted, or retreats furthest into itself. Hence *acfi* . . . is called the *regressive*, or *re-entrant edge*. It is also called the *cuspidal* edge because any section of the surface in a plane cutting *acfi* . . . is of the pointed form $CfC_1$ called a *cusp; Cf* being on the lower and $C_1 f$ on the upper nappe.

The numerous class of surfaces just described, being of infrequent practical use, is sufficiently illustrated by a single example, beginning with a description of its curved directrix.

**100.** *The helix* may be simply defined in two ways, either of which is an obvious consequence of the other..

1°. The helix is generated by a point which has at once a uniform angular motion around a fixed axis, and a uniform rectilinear motion parallel to it. Hence—

2°. *The helix lies on a cylinder of revolution whose axis is that of the helix,* and *crosses its elements at a constant angle,* which is acute.

Hence also, as the elements of a cylinder are parallel both before and after development, *the development of a helix upon that of the cylinder containing it is straight.*

101. In Pl. XIII., Fig. 104, suppose *acfm* to be a helix lying on a vertical cylinder of revolution, and that *ih* is a tangent to it at any point *i*. Then by (52) the angle made by *ih* with the horizontal plane *PQ*, is the angle made at *i* by the helix with that plane. But by (100, 2°) this angle is uniform, hence *all the tangents to a helix so placed, and the helix itself, make a constant angle with the horizontal plane.*

102. Hence (100) the rectilinear development of the helix coincides with its tangent, and the two paths from *i* to *PQ*, *ica* the helical one, and *ih* its tangent, both being of the same constant inclination to *PQ*, are equal. That is—

*The helical arc and its tangent, included between any two parallel planes perpendicular to its axis, are equal.*

### A—Projections.

PROBLEM LXVII.—*To construct the projections of a helix and of its tangent.*

*In Space.*—Both parts of the construction follow directly from the previous definitions.

*In Projection.*—Pl. XIII., Figs. 105, 106. Let the point oo′ in H ascend a distance o′,8′ = $O'O''$ while making a complete revolution around the axis $O$—$O'O''$. Both motions being uniform, divide the circle o8, representing the angular motion, into any convenient number of equal parts, and the corresponding ascent $O'O''$ into the same number of equal parts, and draw horizontal lines through the latter points of division. Then beginning with oo′ the point 1 is projected at 1′ in the horizontal plane through $a_1$, the point 2 at 2′ on the next equidistant line, 3 at 3′ on $a_1$ 3′, etc.

*Otherwise:* making $A$4, Fig. 106, equal to the semicircle o24, Fig. 105, and $A$8 equal to half of $O'O''$, then (100) the straight line 4,8 will be the development of the half turn of the helix from 4,4' to 8,8', Fig. 105. Then divide 4,8 into as many equal parts as there are on the semicircle 468, and draw horizontal lines 55', 66', etc., through the points of division, which will meet the perpendiculars to *GL*, similarly numbered in Fig. 105, at the points 5',6', etc., of the vertical projection of the helix.

*The tangent.*—By (101) the horizontal projections of equal distances on the helix and its tangent will be equal. Hence if *TT'* be a given point of contact, make *Tt* equal to the arc *T*o, and *t* will be the horizontal trace of the tangent at *TT'*. Hence *t* is vertically projected at *t'* and *Tt* and *T't'* are the projections of the tangent at *TT'*.

103. *The developable helicoid* is generated by a straight line which moves so as to be always tangent to a helix, and hence so as to make a constant angle with any plane perpendicular to the axis of the helix.

**PROBLEM LXVIII.**—*To construct the projections of a developable helicoid, its axis being vertical.*

*In Space.*—*First case.* If the surface be given by its helical directrix, its elements can be constructed as tangents to that curve, as in the last problem.

*Second case.* If the surface be given by its axis and one element, any number of additional elements can be found on the principle that their inclination to **H** is constant, and that, therefore, the portions of them included between any two horizontal planes will be equal, as well as equidistant from the axis.

*In Projection.*—Pl. XIV., Fig. 107. *First.* Let *acg*—*a'c'g'Q'* be the *given helical directrix*, constructed as in the last problem. Then, as in the same problem, construct any desired number of tangents to this helix, and they will collectively constitute the projections of the surface. Thus making *dT*, tangent at *d* and equal to the arc *dba*, projecting *T* upon the ground line at *T'*, and drawing *T'd'*, we have $TdD_1$—$T'd'D_1'$ an element of the surface.

*Second.* Let $O$—$O'Q'$ be the axis of the surface, and $aQ''$—$a'Q'''$ the portion of an element included between **H** and the horizontal plane $Q'Q'''$. Then draw various radii $Ob$—$o_ib'$, $Oc$—$o_ic'$, etc.; also $b_ibB_i$, perpendicular to $Ob$ at $b$, and equal to $aQ''$, with $bb_i$ equal to the arc $ba$; $c_icC_i$ perpendicular to $Oc$ at $c$, and equal to $aQ''$ with $cc_i$ equal to the arc $ca$, etc. Finally, project $a$ at $a'$ and $Q''$ at $Q'''$; $b_i$ at $b_i'$ and $B_i$ at $B_i'$, etc., and $aQ''$—$a'Q'''$, $b_iB_i$—$b_i'B_i'$, $c_iC_i$—$c_i'C_i'$, etc., will be elements, identical with those before found at the same points $aa'$, $bb'$, etc. The projections of the helical re-entrant edge can then be sketched, tangent to the successive elements.

The curves $ab_ic_iT$, and $Q''B_iC_i \ldots G_ia$, found by joining the traces of the equal elements just found, on the planes **H** and $Q'Q''$ respectively, are the traces or bases of the helicoid on these planes.

*Visibility.*—*In horizontal projection.* $Q''a$ is visible. Supposing the figure wholly in the *first angle*, $B_ib_i$, for example, is visible from $B_i$ to $b$, and thence invisible till it reappears from $b_i$ to $aQ''$, for below $bb'$ this element is on the lower nappe (99) of the surface, and is therefore concealed by the portion of the upper nappe which is shown.

*In vertical projection.* Situated as the surface is, the inner surface of the upper nappe is seen, except where composed of the parts of elements upward from their point of contact with $gha$—$g'h'Q'$. Also elements of the lower nappe whose horizontal traces accordingly are on $ab_ic_iT$, are visible in vertical projection from the ground line up to their apparent intersection with $a'Q'''$, until we pass the point of contact of a tangent to the base, perpendicular to the ground line.

EXAMPLES.—1°. Construct the figure when turned 90° about its axis $O$—$O'Q'$.

Ex. 2°. Construct the figure when likewise turned 180°. [In these examples certain elements, readily determined by inspection, will be visible throughout. The elements generally will be seen from their visible extremity, readily found, to their point of contact with the helix.]

*Examination of the horizontal trace of the helicoid.*

**104. Involutes. Evolutes.**—When a straight line rolls tangentially upon any curve whatever, any one point of the rolling line will generate a curve which is called an *involute* of the given curve. Thus in Pl. XIV., Fig. 107, when the tangent $aQ''$, considered as a line in **H**, becomes successively tangent at $b$, $c$, etc., and all the intermediate consecutive points, the point $a$ will describe the involute $a\,b_,c_,$, etc., and the point $Q''$ will describe the involute $Q''B_,C_,D_,$, etc.

The fixed curve on which a given straight line rolls is called an *evolute*, relative to its involute.

**105. Tangents. Normals.**—The point $c_,$, for example, is, at the instant of passing that point, describing a circular arc about $c$ the corresponding point of contact of $c_,c$ with the circle $Oa$, hence a *tangent* at $c_,$ will be perpendicular to $cc_,$. That is, *the tangent to the involute at any point is perpendicular to the position of the generating line at that point.*

But the perpendicular as $cc_,$ to a tangent at its point of contact is a normal. That is, *the normal to the involute is a tangent to the evolute.*

**106. Involute of the helix.**—The tangential rolling of $aQ''$—$a's$ upon the circle $Oa_,$ and of $aQ''$—$a'Q'''$ upon the helix $abc$, are simultaneous, being effected by the rolling of the triangle $a'sQ'''$ upon the vertical cylinder whose base is $abeg$. Hence $a$, as the point common to $a'Q'''$ rolling on the helix, and $aQ''$ rolling on the circle $abeg$, describes the same curve $ab_,c_,$ . . . which is thus an involute of the helix as well as of its horizontal projection $abeg$.

## B—Tangencies.

**107.** Tangent planes to cylinders and cones of transposition exist under the same conditions (46), and are constructed in the same way that has already been shown (Prob. XXVIII., etc.)

# DESCRIPTIVE GEOMETRY.

The helicoid above described being developable, tangent planes may be drawn to it under the same conditions as for cylinders and cones, as will next be shown.

**PROBLEM LXIX.**—*To construct a tangent plane along a given element of a developable helicoid.*

*In Space.*—The required plane will be determined by any two tangent lines in it, one of which may be the given element (44).

*In Projection.*—Pl. XIV., Fig. 107. Let $Td$—$T'd'$ be the given element. Knowing the base $ab_,c_,\ldots$ of the surface, $TS$ tangent to it at $T$, and hence perpendicular to $Td$ (105), is at once the horizontal trace of the tangent plane, as well as one of the tangents which determine that plane. The vertical trace of the plane is then determined by the point $S$, and by the vertical trace of $Td$—$T'd'$, or of any tangent parallel to $TS$, as $ij$—$i'j'$, whose vertical trace, $j'$, gives $Sj'$ that of the plane.

EXAMPLE.—Construct the tangent plane on a given element of the developable helicoid situated as in the examples to the last problem.

**PROBLEM LXX.**—*To construct a tangent plane to a developable helicoid, through a given exterior point.*

*In Space.*—Make the given point the vertex of a cone of revolution having the same declivity that elements of the helicoid have. Then a common tangent to the bases of the two surfaces will be the horizontal trace of the required plane. Its vertical trace can then be determined as in the last problem.

*In Projection.*—The construction is left as an example.

**PROBLEM LXXI.**—*To construct a tangent plane to a developable helicoid, and parallel to a given line.*

*In Space.*—Make any point of the given line the vertex of an auxiliary cone of revolution as in the last problem.

Then the required plane will be parallel to one through the given line and tangent to this cone.

*In Projection.*—Pl. XIV., Fig. 107. Let $kl-k'l'$ be the given line. At any point $kk'$ of this line, draw $km-k'm'$ parallel to $aQ''-a'Q'''$, and draw the circle of radius $km$, which will be the base of the auxiliary cone employed. Then $lo$, tangent to the circle $km$, from $l$ the horizontal trace of $kl-k'l'$, will be the horizontal trace of the plane parallel to the required one; hence $UV$ parallel to $lo$ and tangent to the base $ab,c_1$ of the helicoid is the horizontal trace of the required plane. Its vertical trace can then be found as $Sj'$ was, or may be drawn parallel to the vertical trace, also easily found, of the auxiliary plane $lo$.

108. *Declivity of the developable helicoid.*—The element of contact of any tangent plane to this surface, being perpendicular to its horizontal trace, is a line of declivity (Prob. VII.) of the plane. But all the elements, and hence all the tangent planes, have equal declivities; hence the surface is one of the class called surfaces of uniform declivity.

109. *Other examples of this class of surfaces* are those in which the horizontal traces of the tangent planes, all having equal declivities, are tangent to some given curve in the horizontal plane. Thus an embankment of uniform slope surrounding an elliptical or other pond would be a surface of uniform declivity, and so, on a smaller scale, would be the bevelled edge of an oval picture-frame.

110. *Cone director.*—When every element of a surface is parallel, each to some element of a certain cone, the latter is called the cone director of the surface. When, moreover, the tangent plane on any element of the surface is parallel to that on the *parallel* element of the cone director the surface is *developable*. For, see Pl. XIII., Fig. 104, if at any one point lines be drawn parallel to $bc$, $ef$, $hi$, etc., they will form a pyramid having faces, which, with the planes containing them, are parallel to the faces $bdc$, $egh$, etc., of the develop-

able polyedron and *their* planes, Fig. 104. Then as these faces become infinitely many and small, they become elements of a cone director, parallel to corresponding ones of the developable surface, with parallel tangent planes along the parallel elements of the two surfaces.

The cone director of a surface of *uniform* declivity is evidently one of *revolution*.

### C—Intersections.

*Developable Helicoids.*

111. The general, or re-entrant developable surface just illustrated by the developable helicoid, is easily represented by its elements. Its intersection by any plane is, therefore, readily found by constructing the intersections of its elements with that plane by Prob. XI.

112. The intersection of the developable helicoid (Pl. XIV., Fig. 107) with any plane perpendicular to its axis, is a pair of involutes of the circle $abfg$, one on each nappe, and beginning at the intersection of the plane with the helical directrix.

113. The intersection of the developable helicoid with a cylinder having the same axis as the helicoid, consists of two helices, one on each nappe, and projected on H, the common axis being vertical, in a circle concentric with the horizontal projection of the helical directrix.

114. The intersection of any form of the general developable surface with any cone is found by means of auxiliary planes, each of which contains an element of the developable surface and the vertex, and hence, generally, an element also of the cone, whence the rest of the solution is obvious.

115. The intersection of the surface with any cylinder requires the use of auxiliary planes, each of which shall

contain an element of the surface, and shall be parallel to the axis of the cylinder (Prob. IX., 4°).

EXAMPLES.—1°. Find the intersection of the developable helicoid with any plane oblique to its axis.

Ex. 2°. Find the intersection of the same surface with any cylinder.

Ex. 3ᵇ. Find the intersection of the same surface with any cone.

Ex. 4°. In each of the last three examples, construct a tangent line at a given point of the intersection.

*General View of the Intersections of Cones—Infinite Branches.*

116. The manner of locating the auxiliary planes *in each of the three general cases*—two cylinders (59), two cones (Prob. XXXVII.), and a cylinder and a cone (58), and the details of the construction of the points of intersection (57), being essentially the same *in all cases*, the following remarks will be a sufficient guide to the construction of every variety of form which the general problem can present.

117. I. *Varieties of intersection depending on the relation of the tangent auxiliary planes.*

1°. When *both* of the auxiliary planes tangent to *one* body cut the *other* body, the intersection of the two bodies will consist of two separate curves, one of entrance and one of departure.

2°. When *one* tangent plane to *each* body cuts the *other* body, there will be one curve. See Pl. IX., Fig. 78.

3°. When there is *one common tangent plane*, the curve will cross itself *once*.

4°. When there are *two common tangent planes*, the curve will cross itself *twice*, and will consist of a pair of conic sections whenever any plane section of each body is a conic section.

118. II. *Varieties of intersection depending on the relation of either cone, V, to a cone C composed of elements through the vertex of V and parallel to those of the other cone, $V_1$.*

This test cone, *C*, is precisely analogous in its uses to the *plane* through the vertex, whose position determined the character of the parallel *plane* intersections (36).

1°. When the parallel cone, $C$, has *only its vertex* in common with the given cone $V$, the two given cones can have no element of either parallel to elements of the other; hence their intersection can have no infinite branch, and consequently will be a *closed curve*. This, though of double curvature (49), yet being analogous to the closed plane intersection, may be distinguished as belonging to the class of *elliptic intersections*.

2°. When the parallel cone, $C$, is tangent to the same given cone $V$, there will be a pair of parallel elements, $e$ and $e_1$, on $V$ and $V_1$ respectively, and the intersection will consist of one infinite branch having no asymptote (51), since the tangent planes along $e$ and $e_1$ being parallel, their intersection, which (Prob. XXXIX.) is the tangent to the intersection of the cones at the point where the elements $e$ and $e_1$ meet, is wholly at infinity.

The curve in this case, therefore, belongs to the class of *parabolic intersections*.

3°. When the parallel cone, $C$, intersects cone $V$ in two elements, there will be two pairs of parallel elements, as in Pl. XIV., Fig. 111, and two asymptotes (51), and the curve will have an infinite branch with asymptotes, belonging therefore, by analogy, to the class of *hyperbolic intersections*.

4°. When one of the cones becomes a cylinder, the condition for the existence of an infinite branch in the intersection is, that one element of the cone must be parallel to those of the cylinder, and hence must coincide with the auxiliary line through the vertex of the cone parallel to the axis of the cylinder (58). Then, as will be evident on making the construction, that one of the auxiliary planes, all passing through this line (58), which is tangent to the cone, will either not cut the cylinder or will be tangent to it, or will cut it in two elements. In the *first* case, it will give no points, being outside of the limiting planes (Prob. XXXVII., 3°) which include the intersection. In the *second* case, there will be one infinite branch, and in the *third*, two of them.

After the construction already given (Probs. XXXVII., etc.) the foregoing principles are sufficiently illustrated in the next problem.

**PROBLEM LXXII.**—*To find the intersection of two cones when the curve has infinite branches.*

*In Space.*—Adjust the form or position of the given cones until by applying (118) it is found that there will be infinite branches, and then proceed as in Prob. XXXVII.

*In Projection.*—Pl. XIV., Fig. 111. Let $vABk$—$v'A'B'$ and $VCbD$—$V''C'D'$ be the given cones. $vq$—$v'q'$, parallel to the axis $VQ$—$V'Q'$, is the axis, and $vr$—$v'r'$ (where $v'r'$ is parallel to $V''D'$) an extreme element of the auxiliary cone at $vv'$ and similar and parallel to the cone $VV''$. Its base $arh$ intersects $AdB$, that of cone $vv'$, at $a$ and $h$, which shows that two elements of the cone $VV''$ are parallel to the elements $av$ and $hv$ of the cone $vv'$. Then find $X$, the horizontal trace of the line $Vv$—$V''v'$, common to all the auxiliary planes (Prob. XXXVII., *In Space*), and $Xad$ and $Xhk$ are the horizontal traces of the planes of pairs of parallel elements on the two cones. That is, $cV$—$c'V''$ is parallel to $av$—$a'v'$, and $jV$—$j'V'$ to $hv$—$hv''$.

These planes also contain the elements whose horizontal traces are $b$ and $d$ in the plane $Xd$, and $i$ and $k$ in the plane $Xk$, and therefore, as usual, each gives four points of the curve. Thus $va$ and $Vb$ give $e$; $vd$ and $Vb$ give $ff'$; $vd$ and $Vc$ give $gg'$; and finally $va$ and $Vc$, the parallel elements, intersect only at infinity *in each direction* on the infinite branch of the required curve.

Likewise the plane $Xk$ gives the points whose horizontal projections are $s$, $y$, $l$, and the point at infinity, intersection of $hv$—$h'v'$ and $jV$—$j'V'$ in both directions.

*Number and character of the curves.*—As both of the auxiliary planes, $Xx$ and $Xz$, tangent to cone $VV''$, are secant to cone $vv'$, there will be two curves, one at which the smaller cone $VV'$ enters, and one at which it leaves the larger cone $vv'$. One of these being composed of points at which *all* the elements of cone $VV'$, beginning at $w$, intersect those of cone $vv'$ met in proceeding from $z$ to $x$ and back again, is a closed curve. The other, embracing the intersections of all the elements of cone $VV'$ with those of $vv'$ from $v$ to $u$ and back, includes the intersections of the parallel elements, and is therefore infinite. Other relative positions of the cones might have made both curves infinite.

*The tangents to the points at infinity.*—These are found as in any other case. Thus the asymptote (118, 3°) $tm-t'm'$, parallel to the parallel elements $av-a'v'$ and $Vc-V'c'$, is the intersection of the tangent planes $at$ and $ct$ along these elements. The other asymptote $Tn$ is similarly found.

When the parallel elements are in a limiting plane, as $Xx$, the element of contact of that plane itself becomes the asymptote.

*Connection of points.*—(Prob. XXXVII., 2°.) The only peculiarity under this head is the manner of proceeding on passing a plane, as $Xd$, containing parallel elements. The parallel elements $va-v'a'$ and $Vc-V'c'$ must be considered as meeting indifferently in either of their two opposite directions, from $v$ towards $a$ or the reverse; yet, as a point of the curve, and because two straight lines can only meet in one point, these two infinitely distant points must be considered as one point. This is clearly seen by taking auxiliary planes very near to $Xd$, but on opposite sides of it. Call $Y_1$ and $Y_2$ such planes, and let $Y_1$ be behind $Xd$—that is, on the same side with $Xx$. The point $y_1$, then determined in $Y_1$, by the elements near $a$ and $c$, will be very far *behind* $Xd$. Next $Xd$ itself will give the point $x_1$ at infinity. Then in $Y_2$, the elements near and in front of $a$ and $c$ will intersect at $y_2$ very far *in front* of $Xd$. This shows that the curve passing successively, *without reversing its direction*, through $y_1$, $x_1$, $y_2$, the point $x_1$ must be considered indifferently as infinitely far before and behind $Xd$.

EXAMPLES.—1°. Make $Xd$ coincide with $Xx$ by changing the vertices so that $uv$ and $pV$ shall be parallel.

Ex. 2°. Let the given cones be so placed that *one* only of the tangent auxiliary planes to each shall be secant to the other.

Ex. 3°. Place the cones so that the auxiliary cone shall be *tangent* to one of the given cones.

Ex. 4°. Replace one of the cones by a cylinder as in (58) and find the curve.

[Devote a whole plate to Prob. LXXII., and to each of these examples, making a complete study.]

## D—Development.

119. The direct development of a cone of transposition consists in developing an inscribed pyramid of many sides. This method is inexact, both in principle and mechanically, and the construction is therefore postponed until, by means of a combination of developable and double-curved surfaces, a better practical solution shall be found.

PROBLEM LXXIII.—*To develop the convex surface of a cylinder of transposition whose elements are oblique to both planes of projection.*

*In Space.*—The solution embraces these three distinct operations.

1°. Find the intersection of the cylinder with a plane perpendicular to its axis by (Prob. XXXIV.), since only such a section will develop into a straight line showing the circumference of the right section.

2°. Find the true size of this right section by revolving its plane into or parallel to **H** or **V** as in (Prob. XL.)

3°. Find the true lengths of the segments of elements between the right section and the base or any other given section, as in (Prob. XII.)

*In Projection.*—1°. Lay off on a straight line a distance equal to the circumference of the true size of the right section, dividing the latter for this purpose into a sufficient number of parts, $a$, $b$, $c$, etc.

2°. Find, as described, the true lengths of the elements

through these points of division, and make them all perpendicular to the developed right section at the corresponding points $a_1$, $b_1$, $c_1$, etc., and connect their extremities, which will give the developed circumference of the base or other limiting section chosen (33).

As the construction requires no new operations, it is left as an example.

PROBLEM LXXIV.—*To develop a helix by means of its curvature.*

*In Space.*—Since both of the component motions of the generatrix of a helix are uniform, the rate of curvature of the curve must be so also, hence when this rate, as shown by the radius of curvature, is known, the helix can be developed as a circular arc.

*In Projection.*—See first Pl. XIV., Fig. 108. Let $ee'$, $ff'$, $nn'$ be three equidistant points on a helix, and let the plane $jbj'$ be perpendicular to both **H** and **V**. Suppose a circle to be passed through these three points, $e'f'n'$ is on its vertical projection, and its diameter, vertically projected at $f'$, is the hypothenuse of a right-angled triangle in which $fn$—$f'n'$ is one side. Drawing the chord $en$—$e'n'$, denoting the points in space by the capitals of the same letters used on the figure, and calling $D$ the diameter of the circle $EFN$, we have

$$(FN)^2 = D \times fo.$$

Likewise in the triangle $fnd$, calling $fd = 2r$,

$$(fn)^2 = 2r \times fo.$$

Divide these equations member by member, and let $n'e'c' = \beta$, and

$$\frac{1}{\cos^2\beta} = \frac{D}{2r}$$

Now as $N$ successively approaches and finally coincides with $F$, the angle $\beta$ approximates and at last equals $a$, the angle made by the tangent at $F$ with $\mathbf{H}$, and at the same time $D$ approaches and finally becomes $2\rho$ the diameter of the osculatory circle containing $E$, $F$ and $N$ when these points become consecutive, and whose radius is therefore the radius of curvature of the helix; hence, finally,

$$\rho = \frac{r}{\cos^2 \alpha}$$

This value is easily constructed as follows: Fig. 107. Lay off $Q'''p = Oc = r$; draw $pq$ perpendicular to $Q'Q'''$; draw $qr$ perpendicular to $a'Q'''$, and $Q'''r$ will be the value of $\rho$. For

$$r = Q'''p = Q'''q \cos \alpha, \text{ or } Q'''q = \frac{r}{\cos \alpha}$$

but $\qquad Q'''p : Q'''q :: Q'''q : Q'''r$

hence $\qquad Q'''r = \rho = \dfrac{r}{\cos^2 \alpha}$

Therefore, finally, an arc of radius $Q'''r$ and length equal to $Q'''a'$ will be the development of the helix $acg$—$a'c'g'Q'$, Fig. 107, by means of its curvature.

**PROBLEM LXXV.**—*To develop so much of the surface of a developable helicoid as lies between two planes perpendicular to its axis.*

*In Space.*—The helix is developed by means of its curvature, since the elements tangent to it in space must, to preserve their relative position, be tangent to its equivalent plane curve (60) of the same curvature (98). Also the required portions of the elements being equal, will be so in development, and of their true length.

*In Projection.*—To develop the helicoid, Fig. 107. Pl. XIV., Fig., 109. Describe a circle $Oc$ of a radius equal to $Q'''r$, Fig. 107. On it lay off the arc $acgq$, equal to $Q'''a'$, Fig. 107, which will, by the last problem, be the proper development of the helix in Fig. 107. At the points $a, b, c, d$, etc., of this arc, corresponding to $aa'$, $bb'$, $cc'$, etc., Fig. 107, draw the tangents $aQ$, $a_1bQ_1$, $a_2cQ_2$, etc., each equal to $Q'''a'$, the true length of an element, and placed as in Fig. 107—that is, so that the same point of the helix and of a tangent shall coincide in space and in development. Then will $a_1b = Arcab$; $a_2c = abc$; $Q_2c = qfdc$, etc., making the developments of the involute bases, involutes of the developed helix beginning at $a$ and $q$.

*The shortest distance* between two points in a plane being straight, the development of the shortest path on the helicoid between any two of its points will be a straight line joining the developments of those points. From this development, the projections of the same line can easily be constructed.

120. *Examination of consecutive elements.*—Several examples of the apparent contradictions that arise whenever an infinity is involved have already occurred: The question whether a tangent, as *requiring two points to determine it*, shall be defined as containing two consecutive points, or, *not being a secant*, as containing only one point of a curve (52): That of the parabola as expanding in width to infinity, yet as necessarily cutting the elements consecutive to the one to which it is parallel on the cone containing it (36, 2°): That of connecting the points of infinite branches of intersections where one point of the curve is, in construction, either of two points at an infinite distance apart (Prob. LXXII.)

121. A new and interesting illustration of the last article occurs in connection with the developable helicoid. Defined as the limit reached when the edges of a developable polyedron, Pl. XIII., Fig. 104, become consecutive, it is plainly developable. Yet tangents to a curve of double curvature (97) are not in the same plane.

But see now Pl. XIV., Fig. 110, where $ab$—$a'b'$ is a curve of double curvature placed in the simplest position—that is, so that the tangents at $aa'$ and $bb'$ shall be parallel to **H**, while one of them, that at $aa'$, is perpendicular to **V**.

The line $n$—$a'n'$ is the common perpendicular, or *shortest distance* between these tangents. Then, drawing the chord $ab$—$a'b'$, we have

$$a'n' = b'c' = a'c'.\, tang.\, b'a'c' = ad.\, tang.\, bad.\, tang.\, b'a'c'.$$

Now as the point $bb'$, continuing on the curve, approaches $aa'$, the plane **H** must rotate about the tangent $ad$—$a'$ in order to remain parallel to the tangents at $aa'$ and the moving point $bb'$, and the limit of its successive positions will be the osculatory plane containing the three consecutive points of which $aa'$ is the middle one. Also the plane $aa'b'$ of the tangent at $aa'$ and the point $bb'$ will, in the same approach of $b$ to $a$, have the same limit. Consequently the angle $b'a'c'$ and the distance $a'n'$, and hence all the factors of the above equation, become infinitely small together. Therefore $a'n'$ becomes an infinitesimal of the third order—that is, as each of the factors is then an infinitely small fraction of any finite unit, their product is an infinitesimal part of an infinitesimal. Hence when infinitesimals of different orders occur together, the higher can be neglected in comparison with the lower, but not with respect to each other; as ordinary ones are, as compared with finite quantities. Accordingly, we may say, for all purposes involving finite quantities, that consecutive tangents to a curve of double curvature do intersect.

## CHAPTER III.

### WARPED SURFACES OF TRANSPOSITION.

*Principles.*

**122.** WARPED surfaces of transposition are of endless variety, depending on the conditions to which the motion of the generatrix is subject. But these conditions may all be reduced to the two following. A warped surface may be generated—

1°. By a straight line, $G$, moving upon three fixed lines $C, C_1, C_2$; either of them straight or curved.

2°. By a straight line, $G$, moving on two fixed lines and parallel to a cone, called the *cone director*.

**123.** *The first method.*—If any point $c$ of a curve $C$ be made the vertex of a cone whose directrix is $C_1$, this cone will, in general, cut any portion of $C_2$ in only one point, as $c_2$. Then the line $cc_2$, resting on $C, C_1, C_2$, will be fixed. But proceeding likewise with a point $c'$, consecutive with $c$ on the curve $C$, the new element $c'c_2'$, consecutive with $cc_2$, could not generally be in the same plane with $cc_2$ when the three curves $C, C_1, C_2$ were arbitrarily chosen. Hence,

*A warped surface may be generated by a line moving on three given fixed lines.*

Supposing then any warped surface to be given, generated under any conditions whatever. *Any three curves traced upon it may be considered as the curves $C, C_1, C_2$, by which, as directrices, it might have been generated.*

If the three fixed lines are straight, the cone described will become a *plane* (64).

**124. *The second method.*—**If through any point $p$ in space, lines be drawn parallel to all the elements of a warped surface, they will constitute a cone to which all the elements will be parallel, and which is called the *cone director.*

Now, if any point $c$, of any curve $C$, be made the vertex of a cone $V$ similar to a fixed given cone $V_0$, the cone $V$ will in general cut any assumed curve $C_1$ in only one point $c_1$. Hence the line $cc_1$, resting on the curves $C$ and $C_1$, and parallel to some element of the cone $V_0$, is fixed. As before, the consecutive line $c'c_1'$ could not generally be in the same plane with $cc_1$. Hence—

*A warped surface may be generated by the second method; and, however actually generated, in any given case,* any two curves, $C$ and $C_1$, traced upon it, and a cone having elements parallel to its elements, may be taken as its two directrices and cone director.

**125. *The simplest case*** under the *first* method, is when *all the three directrices are straight*, and in particular at equal perpendicular distances, taken in one plane, from a fourth line. This gives the *warped hyperboloid of revolution.* (Prob. XLIII.)

*The simplest case* under the *second* method is when both directrices are straight, and the cone director reduces to a plane, called the plane director. This gives the surface called the *hyperbolic paraboloid.*

**126.** We shall illustrate warped surfaces of transposition by the following examples, taking those which differ most from each other, and are of most frequent use:

1. Hyperbolic paraboloids. 2. Elliptic hyperboloids. 3. Helicoids. 4. Conoids. 5. Warped arches. 6. General warped surfaces.

For convenience of comparison, the first five of these are here defined together.

**127.** 1°. In the *hyperbolic paraboloid*, the generatrix moves upon two fixed straight lines, its *directrices*, and is always parallel to a fixed plane, its *plane director.*

2°. The *elliptic hyperboloid* differs from that of revolution (66) in that its plane sections, perpendicular to the axis, are similar ellipses.

3°. The *warped helicoid* is generated by a straight line moving upon a helix and its axis as *directrices*, and making a constant angle with the axis.

4°. The conoid has two directrices and a plane director, but one of the directrices is curved.

5°. The *warped arch* has three directrices, viz., *two circles* in parallel planes, and *a straight line*, which, together with the line joining the centres of the circles, is in a plane perpendicular to those of the circles.

THEOREM X.—*The hyperbolic paraboloid is doubly ruled, or has two sets of elements.*

Pl. XIV., Fig. 112. Let $Ka$ and $Mb$ be two lines not in the same plane, and let $ab$, $DE$, $KM$, be three positions of a line moving on $Ka$ and $Mb$, and always parallel to the plane $HP$. The moving line will then generate a hyperbolic paraboloid.

Now, let $V$ be a plane, parallel to $Ka$ and $Mb$, and cutting $ab$, $DE$, $KM$ at $c$, $F$, $N$, and let $Mbm$ and $Kak$ be planes parallel to $V$, through the lines $Mb$ and $Ka$. These three planes will then cut $HP$ in the parallel lines $GL$, $bm$ and $ak$.

Again: drawing $Ff$ and $Nn$ perpendicular to $GL$, pass planes through $DF$ and $Ff$ and through $KN$ and $Nn$. They will cut $HP$ in the lines $df$ and $kn$, parallel to $DF$ and $KN$, and the planes $Mbm$ and $Kak$ in the lines $Ee$ and $Dd$ equal and parallel to $Ff$, and $Mm$ and $Kk$ equal and parallel to $Nn$.

We shall now show that $NFc$ is a straight line.

The triangles $Kak$ and $Mbm$ give:

$$ad : ak :: Dd : Kk \quad \ldots\ldots\ldots\ldots\ldots\ldots(1)$$
and
$$be : bm :: Ee : Mm \quad \ldots\ldots\ldots\ldots\ldots(2)$$

But by reason of the equality of the third and fourth terms

$$ad : ak :: be : bm \ldots\ldots\ldots\ldots\ldots\ldots(3)$$

which shows that $ac$, $df$, $kn$ all meet in one point, hence also

$$ad : ak :: be : bm :: cf : cn \dots\dots\dots(4)$$

and as $Ff = Dd$ or $Ee$, and $Nn = Kk$ or $Mm$.

$$cf : cn :: Ff : Nn \dots\dots\dots\dots\dots(5)$$

which shows that $NFc$ is a straight line, and hence that the same surface can be generated, either by the motion of $ac$, parallel to $HP$, and upon $Ka$ and $Mb$, or by the motion of $Nc$ parallel to $V$, and upon $NK$ and $FD$.

That is, the hyperbolic paraboloid is doubly ruled, as stated.

128. *Consequences of the double generation.*—The hyperbolic paraboloid is seen by the last theorem to have *two plane directors*, one for each set of elements. These two sets are distinguished as those of the *first and second generation.* Moreover, it is now evident that *any element of either generation intersects all those of the other*, and hence that the directrices are no peculiar lines, but that *any two elements of either generation may be taken as the directrices of the other*.

When the plane directors are perpendicular to each other, the surface is called *right, or isosceles ;* otherwise, it is called *scalene*.

129. *Proportional division of elements.*—Substituting $aD$ and $aK$, $bE$ and $bM$, $cF$ and $cN$ for $ad$, $ak$, etc., in the proportion (4) of Theor. X., we see that the elements of one generation divide those of the other, and hence the directrices, proportionally. Also by drawing perpendiculars to $GL$ from $a$ and $b$, passing planes through them and $aK$ and $bM$ respectively, and passing other planes through $FD$ and $KN$ parallel to $HP$, we can directly demonstrate as before, that

$$ab : bc :: DE : EF :: KM : MN.$$

130. *Vertex. Axis.*—As the elements parallel to $V$ make larger angles with $HP$ as we proceed from $Ka$ towards $V$,

there will be one of them whose direction is perpendicular to that of *GL*. Also as indicated by *ac* and *kn*, some element between *ac* and *KN* will be perpendicular to the direction of *GL*. The intersection, *v*, of these elements, one of each generation, is called the *vertex* of the surface.

A line through *v*, parallel to *GL* the intersection of the plane directors, is called the *axis* of the surface.

### A—Projections.

**PROBLEM LXXVI.**—*To construct the projections of a hyperbolic paraboloid, having given the directrices and plane director of one generation.*

*In Space.*—The line joining the points in which the plane director cuts the directrices will be one element. The line joining the points in which a parallel plane cuts the same directrices will be a second element.

Having thus two elements, apply the principle of (129), and lay off on each directrix parts equal to those intercepted on it between the parallel planes, and the lines joining the points of division will be elements. Or, for better accuracy, let the parallel planes be as far apart as possible, and divide the space between them on each directrix into the same number of equal parts.

*To construct elements of the second generation.*—Pass a plane by (Prob. IX., 4°) through one of the given directrices and parallel to the other. This will be the plane director of the second generation, and any two of the elements before constructed will be its directrices; and we can then proceed as in constructing the elements of the first generation.

*In Projection.*—Pl. XIV., Fig. 113. Let $AB$—$A'B'$ and $CD$—$C'D'$ be the directrices, and $PQP'$ the plane director. By (Prob. XI., *a*) the plane $PQP'$ cuts $AB$—$A'B'$ at $oo'$, and $CD$—$C'D'$ at $oo'$. Then $oo$—$o'o'$ is an element of the required surface. $RS$, parallel to $PQ$, is the horizontal trace of a plane parallel to $PQP'$ as shown by making $c1$—$c'1'$ and

$gd$—$g'd'$ parallel to $ab$—$a'b'$ and $ef$—$e'f'$, these being the intersections of these planes with the auxiliary vertical planes containing the directrices. Thus the plane $RS$ cuts the directrices at 1,1' and $dd'$, giving $1d$—$1'd'$ for a second element.

Finally, lay off on $AB$—$A'B'$ spaces 12—1'2', etc., each equal to 01—0'1', and on $CD$—$C'D'$, spaces $dl$—$d'l'$, $lq$—$l'q'$, etc., each equal to $od$—$o'd'$, and the lines, as $3q$—$3'q'$, joining the like points, will be other elements, any number of which can thus be constructed.

EXAMPLES.—1°. Vary the figure, to become familiar with the surface by its projections, by taking numerous and widely different positions of the directrices and plane director.

Ex. 2°. Suppose both plane directors to be vertical. That is, assume the directrices so that elements of each generation shall be parallel in horizontal projection.

Ex. 3°. In Pl. XIV., Fig. 113, construct an element of the second generation. [The above solution in space being only a repetition of Prob. IX., 4°, and of the above operations, taking any two of the elements there found as directrices, the construction is made an example.]

PROBLEM LXXVII.—*Having one projection of a point on the surface of a hyperbolic paraboloid, to find its other projection.*

*In Space.*—Pass a plane through that projecting line (Prob. XI., *b*, 1°) of the point, which determines the given projection, and find the intersection of this plane with the surface. The intersection of this line with the given projecting line will be the point whose required projection will thence be known.

*In Projection.*—Pl. XIV., Fig. 113. Let $p$ be the given projection of a point on the paraboloid. A line $pk$ in any convenient direction will then be the horizontal trace of a vertical plane containing the projecting line whose projection is $p$. This plane intersects the elements of the paraboloid, giving the curve $knr$—$k'n'r'$, on which $p$ is vertically projected at $p'$, the point required.

EXAMPLE.—Let $p'$ be given, and find $p$.

131. Four straight lines intersecting at four points, not in the same plane, form a *warped quadrilateral*.

The opposite sides of this quadrilateral, as $A'B'$ and $C'D'$, Pl. XIV., Fig 114 (the vertical projection), are pairs of elements, one pair of the first, and the other of the second generation of a hyperbolic paraboloid, for through $A'B'$, for example, a plane can be passed parallel to $C'D'$, which will be the plane director of these elements. Likewise, a plane can be passed through $A'C'$, parallel to $B'D'$, which will be the plane director of these elements.

**PROBLEM LXXVIII.**—*Having given a hyperbolic paraboloid by a pair of elements of each generation, and having also a point of the surface, to construct an element of either generation through that point.*

*In Space.*—Since all the elements of the same generation are parallel to the same plane, a pair of lines through the given point, parallel to elements of the first generation, will determine a plane through that point and parallel to the plane director of that generation. Hence the intersection of this plane with the elements of the other generation will be points of an element of the first generation.

*In Projection.*—Pl. XIV., Fig. 114. Let $AB$—$A'B'$ and $CD$—$C'D'$ be elements of the first generation, and $AC$—$A'C'$ and $BD$—$B'D'$ those of the other, and let $pp'$ be the given point. In practical cases, $pp'$ may have been found as in the last problem, more elements of either generation being given than are here shown.

Through $pp'$ draw $pa$—$p'a'$ and $pd$—$p'd'$, respectively parallel to $AB$—$A'B'$ and $CD$—$C'D'$, and they will determine a plane containing an element through $pp'$. By (Prob. XI., *b*, 2) this plane cuts $AC$—$A'C'$ at $qq'$, and $BD$—$B'D'$ at $rr'$, hence $qr$—$q'r'$ is an element of the first generation, through $pp'$.

Likewise, $pc$—$p'c'$ parallel to $AC$—$A'C'$, and $pb$—$p'b'$, parallel to *BD—B'D'*, determine the plane containing the element *ns—n's'* of the second generation, through *pp'*.

132. A straight line, moving upon three straight lines which are parallel to a plane, but not to each other, generates a hyperbolic paraboloid. For, let *PQ*, Pl. XV., Fig. 117, be the plane, and *cd, ef, gh*, the three given lines, and let *ga* and *hb* be any two positions of the line which moves upon them. Then *cd, ef, gh* are evidently three elements of the hyperbolic paraboloid of which *ag* and *bh* are directrices and *PQ* the plane director; whence (Theor. X.) *ag* and *bh* are elements of the other generation of the same surface.

PROBLEM LXXIX.—*To construct the projections of a warped elliptical hyperboloid.*

*In Space.*—Parallels of a hyperboloid of revolution (33) appearing in horizontal projection as concentric circles, while those of the elliptical hyperboloid appear as similar and similarly situated concentric ellipses, the appropriate construction of the latter is by (Prob. XXI., 2°) as oblique projections of the former.

*In Projection.*—Pl. XV., Fig. 116. The surface is sufficiently illustrated by the figure, which shows the quarter below the gorge *a'b'*, and in front of a vertical plane *AB*, through the axis *O—C'O'* of the surface.

Let *AB* and *2CO* be the axes of the elliptical base *ACB—A'B'*, and *ab—a'b'* the transverse axis of the gorge. *Oc*, the semi-conjugate axis of the gorge, is then a fourth proportional to *OA, OC*, and *Oa*, found by drawing a line *ac* parallel to *AC*.

As a pair of tangents to the gorge issue in horizontal projection from every point of any other parallel alike for the circular and the elliptic hyperboloid, we construct the ellipse *ACB* by (Prob. XXI., 2°), and *f, g, C, h*, etc., will then be points symmetrically distributed relative to *AB* and *2OC*, and from which the horizontal projections of elements

$fm$ and $fk$, $ga$ and $gB$, etc., will issue, as shown, tangent to the horizontal projection of the gorge.

*The vertical projections of elements.—First method.* By the method of (Prob. XXII.), find the exact points of contact $n$, $r$, etc., of elements, $mC$, $Ah$, etc., and project them as at $n'$ $r'$, etc. Also project $A$, $f$, $g$, $C$, etc., at $A'$, $f'$, $g'$, $C'$, etc. Then the two elements issuing from $AA'$ will be vertically projected in $A'r'$, because they are symmetrical relative to the vertical plane on $AB$. The two issuing from $ff'$ will be projected in $f'n'$ and $f'O'$, the first in $f'n'$, because the two issuing from $mm'$ and coinciding in horizontal projection with $fm$ and $Cm$ are, as there seen, tangent to the gorge at $n$ and at a point symmetrical with it relative to $AB$. The two issuing from $gg'$ are vertically projected at $g'a'$ and $g's'$, and so on.

*Second method.*—Draw a line $A''B''$ equal to $A'B'$ and symmetrical with it relative to $a'b'$, and project $A$, $f$, $g$, $C$, $h$, etc., upon it at $A''$, $f''$, $g''$, etc., as well as upon $A'B'$, as before done. Then join the same letters, one on $A'B'$ and one on $A''B''$, that are joined in horizontal projection. That is, draw $A'h''$ and $A''h'$, whose intersection will be the point $r'r$ of the gorge, etc.

In the figure, $hb$ is perpendicular to $AB$ because $Ob$ is half of $OB$, while $HB$ is an arc of 60°. Any other proportions and divisions of $AHB$ might have been chosen.

EXAMPLES.—1°. Make the projections of the entire surface.

Ex. 2°. Make the projections of the entire surface with $AB$ perpendicular to the ground line.

## PROBLEM LXXX.— *To construct the projections of an oblique helicoid.*

*In Space.*—The construction is founded immediately on the definition, according to which the rectilinear elements join points of equal division on the helical directrix with corresponding points of equal division on the axial directrix.

*In Projection.*—Pl. XV., Fig. 115. Let 0369—o'3'6'9', constructed as in (Prob. LXVII.), be the helical directrix, and $O$—$O'O''$ the axis of an oblique helicoid, and let $oO$—$o'o''$, making any given angle, $o'o''O'$, with the axis, be the initial element, parallel to **V**. Then set off on the axis from $o''$ spaces, equal to those laid off from $O'$ in constructing the helix, to mark the uniform ascent of all points of the generatrix. The horizontal projections of elements are then radial lines $Oo$, $O1$, etc., of the circle 0369, and their vertical projections are $o'o''$, $1'1''$, $2'2''$, etc.

*Contour and visibility.*—The apparent contour in vertical projection is a curve—not shown—tangent to the successive vertical projections of elements.

In horizontal projection, every radial represents the visible parts of one or more elements.

Representing the surface as a warped polyedron, as in the figure, each element beginning at a visible point of it, in **H** readily seen by inspection, continues visible upward, as shown in the figure (supposing elements in the fourth angle visible at $m'n'$ and $n'a'$), till it apparently intersects the next, or disappears at its apparent intersection with the apparent contour.

133. The oblique helicoid has a *cone director* (122) whose axis is that of the helicoid, and whose elements make the same angle with that axis that is made by those of the helicoid.

When the angle just mentioned is one of 90°, the helicoid is called a *right helicoid*, and evidently has a *plane director* perpendicular to its axis.

Besides these helicoids, there is one of a more general form, in which the axis is replaced by a cylinder, separate from that containing the helical directrix, and to which the elements are tangent, while also resting upon the helix. In this case these two cylinders have a common axis, but may be of any form, of revolution or not.

**PROBLEM LXXXI.**—*To construct the projections of a conoid.*

*In Space.*—In practice, the plane director may generally be made a plane of projection, when the elements will be immediately determined by their projections parallel to the ground line, on the other plane of projection, which will then show where they intersect the given directrices.

*In Projection.*—Pl. XV., Fig. 119. Let **H** be the plane director, $A-A'B'$ the rectilinear, and $CD-C'E'D'$ the curved directrix (127, 4°).

Parallels to the ground line $C'F'$, as $G'h'$, $L'k'$, etc., are then the vertical projections of elements, whose horizontal projections are found by projecting $G'$, $L'$, etc., at $G$, $L$, etc., and drawing $GA$, $LA$, etc.

EXAMPLES.—1°. Let the rectilinear directrix be oblique to the plane director.

Ex. 2°. Let the directrices and the plane director have any oblique positions in space. [Take any three or more lines in the plane director by Prob. VII., and draw parallels to them from any point, $p$, on the curved directrix. Then find by (Prob. LXXXIV.) where the plane thus determined cuts the straight directrix, and the line from this point to the assumed point $p$ will be an element.]

134. *Right conoid.*—When the rectilinear directrix of a conoid is perpendicular to the plane director, the surface is called a *right conoid*, and this directrix, the *line of striction*, as containing the shortest distances between the elements. In fact, every warped surface has such a line, but it is generally curved.

The conoid is a very comprehensive form of warped surface, and includes some of the cases of other warped surfaces. Thus a right helicoid is also a form of right conoid by the definition of the latter.

**PROBLEM LXXXII.**—*To construct the projections of a warped arch, also called the Cow's Horn.*

*In Space.*—The plane to which the curved directrices are both parallel (127, 5°) may be taken as a plane of projection, when their intersections with any plane containing the straight directrix will be immediately obvious, and will be the points which determine the element in such a plane.

*In Projection.*—Pl. XV., Fig. 121. Let the semicircles $AB$—$A'D'B'$ and $C'E'$—$C'D'E'$, this one in **V** and the other parallel to it, be the curved directrices, and let $OF$—$F'$, in **H** and perpendicular to **V**, be the straight directrix. Then, at once, any lines, as $Fa'c'$ and $F'f'e'$, radiating from $F'$, and intersecting both semicircles, are vertical projections of elements. Their horizontal projections are found as at $caO$ by projecting $c'$ at $c$ and $a'$ at $a'$ on the horizontal projection of the proper semicircle and joining the points so found.

The element at $D'$ is horizontal; those to the right of it, as $ef$—$e'f'$, meet $OF$ behind **V**, while those to the left of it, as $ac$—$a'c'$, meet $OF$ in front of **V**.

When the distance $A'C'$ becomes nothing, the surface becomes a cylinder. When it equals $C'E'$ by the coincidence of $E'$ with $A'$, the surface reduces to two oblique cones tangent to each other on the line $AA'$.

EXAMPLES.—1°. Let $OF$, still in **H**, be oblique to **V**. [The traces of planes, as $OFm'$, upon the parallel planes $AB$ and $CD'$, instead of coinciding as in $Fm'$, will be parallels from the then separate vertical projections of $D$ and $F$.]

Ex. 2°. Draw the figure with $A'$ to the left of $C'$.

**135.** The most general surface of which the warped arch used in practice is a specific example, would be one having any three directrices, one of them straight, and in any position.

**136.** The foregoing are all the warped surfaces which are of any considerable use in the arts, and nearly all which have received specific names and investigation. One, termed the *cylindroid*, and which may serve as the surface of a descending arch connecting, transversely, two parallel arched passages on different levels, is thus derived. Imagine a cylinder whose elements are *parallel to the ground line*. Intersect it obliquely by two vertical planes lying on opposite sides of the plane of right section, as at $Cm$ and $Cp$, relative to

CO, in Pl. XV., Fig. 116. Elevate one of the oblique sections, A, keeping it in its own plane and so that all its points shall ascend equally to its new position $A'$, the other section B remaining fixed. Then the lines joining the same points of $A'$ and B, that were joined on A and B by the elements of the cylinder, will still be parallel to **V** as a plane director, but will be elements of a warped surface called a *cylindroid*.

*General warped surfaces* not having specific names, and of either of the kinds mentioned in (123, 124) are of occasional use. The following examples illustrate the methods of construction employed in such cases, as well as the character of such warped surfaces in general.

PROBLEM LXXXIII.—*To construct elements of a general warped surface having three given directrices.*

*In Space.*—The construction directly illustrates the definition of such surfaces (123).

*In Projection.*—Pl. XV., Fig. 122. Let $AB$—$A'B'$, $CD$—$C'D'$, $EF$—$E'F'$ be the three directrices. On any one of them, as $AB$—$A'B'$, assume any point as $mm'$, and join it by straight lines with points as $aa'$, $bb'$, $cc'$, taken on either of the other directrices. These lines form a conic surface which intersects the cylinder which projects the remaining directrix $EF$—$E'F'$ on **H** in a curve whose horizontal projection, $def$, coincides with $EF$, and whose vertical projection, $d'e'f'$, is found by projecting $d$, $e$, $f$ upon $m'a'$, $m'b'$, $m'c'$. This curve intersects $EF$—$E'F'$ at $nn'$, which is thus the point where the conic surface cuts $EF$—$E'F$. Hence the line $mn$—$m'n'$ by construction intersects the three directrices, and is evidently the only one through $mm'$ that does so; it is therefore an element of the required warped surface. Other elements can be similarly found.

EXAMPLES.—1°. Let the given point $mm'$ be on either of the other directrices.
Ex. 2°. Let one or more of the directrices be in other angles than the *first*.

## 166 DESCRIPTIVE GEOMETRY.

**PROBLEM LXXXIV.**—*To construct elements of a general warped surface having two directrices and a plane director.*

*In Space.*—Any element may either be made *parallel to a given line in the plane director*, or through a *given point on one of the directrices.*

In the *former case*, it will lie on the surface of a cylinder whose elements are parallel to the given line, and which has one of the directrices for its base. In the *latter case*, the element sought will lie in a plane parallel to the plane director, through the given point.

A determining point of the element will be where the cylinder in the first case, and the plane in the second, cuts the other directrix.

*In Projection.*—Pl. XV., Fig. 123. Let $AB$—$A'B'$ and $CD$—$C'D'$ be the directrices, and $MN$—$M'N'$ the plane director of a warped surface. Then for the *first case*, $ab$—$a'b'$ being a given line in the plane director, draw the elements $ad$—$a'd'$, $be$—$b'e'$, $cf$—$c'f'$ of the auxiliary cylinder parallel to it, and having $AB$—$A'B'$ for its directrix (94), and find, just as in the last problem, the point $nn'$ where this cylinder cuts the other directrix. Then $mn$—$m'n'$, parallel to $ab$—$a'b'$, is the element required.

For the *second case*, had the element been required to pass through $aa'$ for example, we should first have taken several other lines besides $ab$—$a'b'$ in the plane director, and should then have drawn parallels to them through $aa'$. These lines would have formed a plane parallel to $MN$—$M'N'$ and containing $aa'$. Its intersection with $CD$—$C'D'$, found exactly as in the two preceding constructions, would give the point which, joined with $aa'$, would be the element required.

EXAMPLES.—1°. Make the construction just described.
Ex. 2°. Vary the position of the plane director.

### B—Tangencies.

137. *Every tangent plane to a warped surface is in general a secant plane also.* This arises from the property, more or

less obvious on inspection of any of the warped surfaces already shown, that any two sections made by secant planes at right angles to each other, through any point, $p$, of the surface, will generally be curves convex in opposite directions, as shown on the hyperboloid of revolution (Theor. VIII.). Hence the tangents to these curves at $p$ must be on opposite sides of the surface, and hence the tangent plane determined by them must also be a secant plane.

138. *Every tangent plane to a warped surface is in general tangent only at a point of contact.* For as the elements of a warped surface are never, except in the case of the two sets on the hyperbolic paraboloid, parallel to more than one plane, and are generally parallel to none, a plane passed through any element will, in general, be oblique to all the others, and will therefore cut them all in points forming a curve which will intersect the given element at some point $c$. Then this element, and this curve, being considered as two sections of the surface at $c$, the element which is its own tangent (44) and the tangent to the curve at $c$ determine the plane which is tangent to the surface at $c$ and there only. This plane is thus, as well as in (137), shown to be also a secant plane.

139. It follows, from the last article, that *every plane containing an element of a warped surface is a tangent plane to the surface at some point of that element*—viz., at the point where the element intersects the curve cut from the surface by the same plane. Also that as the plane revolves about this element as an axis, its point of contact with the surface shifts along the element.

The exception to the two preceding articles is that of a tangent plane parallel to the plane director, when the latter exists. Such a tangent plane has an *element* instead of a point of contact, and is not also a secant plane.

This is clearly illustrated by Pl. XV., Fig. 119, where there may evidently be a tangent plane along the whole extent of that element, $EA$—$E'A'$, of the conoid, which is furthest from the plane director, and it is not a secant plane, since there are no two sections through any point on $AE$—$A'E'$, such that one would be convex in an opposite direction to the convexity of the other.

**140. *Raccordment.*—**The mutual tangency of two warped surfaces which are in contact along a common element, is often distinguished as their raccordment. They raccord along this element.

*When two warped surfaces have a common tangent plane at each of three points on an element common to both, they are tangent at all points of that element.* For (see Pl. XV., Fig. 120) let $P$, $Q$, $R$ be the points of contact of such common tangent planes on the common element $PR$. Then any secant plane at each of these points will cut from the two warped surfaces a pair of curves tangent to each other, as $Pa$ and $Pb$ at $P$; $Qc$ and $Qd$ at $Q$; $Rc$ and $Rf$ at $R$. Now, as three directrices determine a warped surface, while the two curves of each pair have no points in common but $P$, $Q$, $R$, their points of contact; if $PR$ moves on $Pa$, $Pc$, $Pe$, it will generate a warped surface everywhere separate, except on $PR$, from that generated by the motion of $PR$ on $Pb$, $Pd$, $Pf$, as directrices. Hence, as the two are tangent, and not secant at $P$, $Q$, $R$, they will be so at every point of $PR$.

**141.** When two warped surfaces have a common plane director, common tangent planes at two points on a common element are sufficient to determine the surfaces as mutually tangent all along that element, since two directrices and a plane director determine a warped surface, and the reasoning is otherwise the same as in the last article.

**142.** *Any number either of warped hyperboloids or of hyperbolic paraboloids* may raccord with any given warped surface, along any given element of it.

This property follows from (Theor. VI.) and (132). For (see Pl. XV., Fig. 120) the three pairs of curves at $P$, $Q$, $R$ may be taken in three parallel planes, or they may not.

143. If they *are* so taken, their common tangent lines at P, Q, R will be parallel to one plane, and the motion of PR on these three tangents will generate a hyperbolic paraboloid (132), which will raccord with both of the given warped surfaces along PR. See Fig. 118, where the directrices AB, CD, EF, and hence their tangents, are all parallel to V, hence a raccording hyperbolic paraboloid on BF will be generated by the motion of BF upon these tangents.

144. If the three pairs of curves *are not* in three parallel planes, neither will their common tangent lines be so, and the raccording surface generated by the motion of PR upon these tangents will (Theor. VI.) be a warped hyperboloid. For three straight lines must, either two or all of them, be parallel to some one plane, while the hyperbolic paraboloid and the warped hyperboloid are the only warped surfaces, all of whose directrices are straight.

145. Moreover, in the former case, an infinite number of sets of parallel planes through P, Q, R may be taken, each of which (143) will contain a pair of curves tangent to each other at those points, and consequently having common tangent lines at those points. One of these sets may be perpendicular to the plane director, to which the corresponding positions of the generatrix PR remain parallel in moving on these three tangents. That is, *one* of all the hyperbolic paraboloids of raccordment may have plane directors perpendicular to each other, and thus be (128) a right hyperbolic paraboloid.

146. Owing to the ease with which planes can be drawn, tangent to the warped hyperboloid (Prob. XLVI.) and in the same manner to the hyperbolic paraboloid, tangent planes to other warped surfaces are sometimes most conveniently drawn by substituting for them one of these as a raccording surface along the element containing the given point of contact.

**147.** Since the tangent plane to a warped surface has generally (138) only a point of contact, the principle of (45) applies, and *any two* of all the tangent lines at the point of contact determine the tangent plane at that point. In the two warped surfaces of double generation, the hyperboloid and hyperbolic paraboloid, the *two elements* at the given point, may be taken as these determining tangents (Theor. VIII.). In other warped surfaces, the single element at that point, with any other tangent, will determine the tangent plane (138).

**PROBLEM LXXXV.**—*To construct the tangent plane at a given point of a hyperbolic paraboloid.*

*In Space.*—Construct an element of each generation through the given point; their plane will be the tangent plane required.

*In Projection.*—Pl. XIV., Figs. 113, 114. Let $pp'$ on either figure be the given point. Then, remembering (128) that the directrices of either generation, as $AB$—$A'B'$ and $CD$—$C'D'$, Fig. 113, are elements of the other generation, construct by (Prob. LXXVIII.) an element of each generation through $pp'$, and the plane of these two lines (147) will be the required plane.

EXAMPLES.—1°. Complete the construction for Pl. XIV., Fig. 113.

Ex. 2°. Complete the construction for Pl. XIV., Fig. 114.

Ex. 3°. Let H be the plane director, and let the directrices be parallel to V, and construct a tangent plane at a given point of the surface thus given.

**PROBLEM LXXXVI.**—*To construct the tangent plane at a given point of an elliptical hyperboloid.*

*In Space.*—The plane is determined as in (Prob. XLVI.).

*In Projection.*—Pl. XV., Fig. 116. Let $tt'$ be the given point. Then two tangents, $tn$ and $ts$, from $t$ to the horizontal projection of the gorge, will be the horizontal projections of the determining elements (Prob. XXII., 2°), whose vertical projections are thence found as in (Prob. LXXIX.).

The traces of the required plane are thence readily found.

**PROBLEM LXXXVII.**—*To construct the tangent plane at a given point of a warped surface by the direct method—that is, without an auxiliary raccording surface.*

*In Space.*—The test of convenient applicability for this method will be the easy construction of a curved section through the given point, to which a tangent line can easily be drawn, which (138), with the element through the given point, will determine the required plane. The helicoid fulfils this test. Then—

*In Projection.*—Pl. XV., Fig. 115. Let 2,2′ be the given point. From the definition (127, 3°) of the helicoid, every point of its generatrix describes a helix; and (100) o369—o′3′6′9′ is the helix containing the given point. By (Prob. LXVII.) 2P—2′p′ is the tangent to this helix at 2,2′, which, with the element b2h—b′h′ at 2,2′, determines the tangent plane PQP′ at the same point.

If either or both of these lines are inconveniently situated, apply (27, 9°) as in every such case, and employ any one or more lines, each intersecting both of these primary ones, in finding PQP′.

If the given point of contact were on the axis as at O,2″, the tangent plane would be vertical, for the helix described by such a point is the axis, and hence its tangent is (44) also the axis. Thus bhh′ is the tangent plane at O2″.

The plane PQP′ makes a larger angle with **H** than the element Ob—2″b′ does, since the perpendicular from O2″ to PQ (Prob. VII.) is shorter than Ob—2″b′. But as the ascent of every helix in one revolution about the axis is uniform, while their radii vary from zero to infinity, the angle made by a tangent plane with **H** will vary from 90°, as just shown, to the angle made by the elements with **H**, when the point of contact is at an infinite distance from the axis.

EXAMPLES.—1°. Construct the tangent plane at 3,3'.
Ex. 2°. Construct the tangent plane at *ee'*, and at *tt'* on 6*f*—6'*f*'.

**PROBLEM LXXXVIII.**—*To construct a tangent plane to a warped surface indirectly—that is, by means of an auxiliary raccording surface.*

*In Space.*—This method usually consists in finding an element of each generation of the auxiliary hyperbolic paraboloid, which is tangent to the given surface along the element containing the given point of contact. The latter element is one of the two required ones, which together determine the tangent plane required.

*In Projection.*—This method is appropriately illustrated in connection with the *conoid* and the *warped arch* surfaces, one of which has, and the other has not, a plane director.

1°. *The tangent plane to a conoid.*—Pl. XV., Fig. 119. Let *gg'* be the given point of contact. *AG—h'G'* is the element of the conoid through that point. Then *GF—G'F'* and *A—A'B'*, the tangents, parallel to **V** at *GG'* and *A,h'*, are directrices, and **H** is the plane director, of a hyperbolic paraboloid tangent to the conoid along *AG—h'G'*. For, as is evident by inspection of the vertical projection, *AG—h'G'*, in moving on these directrices, can divide them proportionally (129) only by continuing parallel to **H**. Hence *AF—B'F'* is the horizontal trace, as well as an element of the auxiliary paraboloid, and consequently *gf—g'f'*, parallel to **V**, and having its horizontal trace in *AF*, is an element of the second generation tangent at *gg'*, counting *AG—h'G'* as one of the first generation. Hence the plane *PQP'* of these two elements, *PQ* being parallel to *AG* (27, 5°), is the desired tangent plane at *gg'*.

2°. *The tangent plane to the warped arch.*—Pl. XV., Fig. 121. Let *pp'* be the given point. *Oc—F'c'*, and *OF—F'*, the one an element, and the other a directrix of the surface, are two sections of it at $O, F'$; which, being rectilinear, are their own tangents; therefore $OFc'$, the plane of these two lines, is the tangent plane at the point $O, F'$. Hence $mOk$—$m'F'k'$, in this plane, with $ha$—$h'a'$ and $nF$—$n'c'$, all parallel to **V**, are three tangents, serving (132) as directrices of the auxiliary tangent paraboloid generated by $Oc$—$Fc'$ in moving upon them.

To find, next, another element of the same generation with $Oc$—$F'c'$, pass a plane (64, 123) through the directrix $mk$—$m'k'$, and a point as $hh'$ of the directrix $ha$—$h'a'$. The horizontal trace of this plane is $Oh$, and its vertical trace $gn'$ is parallel to $m'k'$. This trace $gn'$ cuts the third directrix, at $n'n$, giving $n'h'k'$—$nhk$ as the element desired. Hence $p'r'$—$pr$, parallel to **V**, is an element at $pp'$ of the same generation as the directrices (128); and $PQP'$, which is the plane of the elements $OC$—$F'c'$ and $pq$—$p'q'$, at $pp'$, of the auxiliary paraboloid, is the required tangent plane at $pp'$.

148. *The tangent to a general warped surface*, as at $nn'$, in Pl. XV., Fig. 122, is determined, as in the previous cases, by the element through that point, together with the tangent at $nn'$ to any curved section, as $EF$—$E'F'$, of the surface through that point. There is this difference however: the curved section is of unknown properties, and therefore the tangent to it can only be located by the eye, unless some special construction independent of these properties can be made.

149. The following is such a construction, partly shown. Assume points on $EF$, for example, as $f, e, d$, etc., on each side of $n$, and draw secants $nf, ne, nd$, etc. Note the points, $f_1$ on $nf$, $e_1$ on $ne$, etc., produced on the same side of $n$, where these secants intersect a circle of any convenient radius, and centre at $n$. Lay off on $f_1 nf$ and *in the same* direction from $f_1$ *that $f$ is from* $n$, a distance $f_1 f_2$ equal to the chord $nf$. Do the like for each secant, and a curve $f_2 e_2 d_2$ . . . . will be formed, which will cross the arc whose centre is $n$ at a point $n_1$, where $n_1 n$ is the limit between secants on the one

and the other side of the point $n$, hence $n_1n$ is the tangent line at $n$.

The curve $f_2e_2d_2$ . . . . is called a trial curve, or "curve of error." The method by means of such curves is practically as exact as a geometrical construction, and sometimes more convenient.

## C—Intersections.

150. *The intersection of a warped surface with any plane* is found by constructing the intersections of its elements with that plane by Prob. XI., and then joining the points thus ascertained.

If the warped surface has a plane director, the auxiliary planes should be parallel to that plane. Each of them will cut the given plane in a right line, and the warped surface in one or more elements whose intersection with that line will be points of the required curve.

151. *The intersection of a warped surface with any ruled surface*, other than a plane, is found by noting the intersection of pairs of elements, one on each surface and in the same plane.

Thus, if one of the surfaces be a *cone*, each auxiliary plane may contain an element of the given warped surface, and the vertex of the cone (Prob. IX., 2°).

If one of the given surfaces be a *cylinder*, each auxiliary plane may contain an element of the warped surface and be parallel to the axis of the cylinder (Prob. IX., 4°).

152. If both of the given surfaces be *warped*, the simplest solution for each case must be sought. Thus, if both surfaces have the same plane director, any auxiliary plane parallel to it will contain an element of each surface, whose intersection will be a point of the intersection of the two surfaces. But, in general, it may be necessary to pass a

plane through each of a number of elements of one of the warped surfaces, and thence to note where each element pierces the curve cut from the other warped surface by this plane.

153. The plane sections, not straight, of a hyperbolic paraboloid, are *parabolas*, when made by planes parallel to the intersection of the plane directors, and *hyperbolas* in all other cases.

154. The intersection of a helicoid with any cylinder of revolution having the same axis as the helicoid is a helix.

THEOREM XI.—*The intersection of a right conoid having an elliptical directrix, with any plane parallel to that directrix, is an ellipse.*

In Pl. XV., Fig. 119, $CD—C'E'D'$ is a semicircle, and the directrix of the conoid there shown, and $cd—c'e'd'$ is the section made by a plane parallel to $CD$. Then by similar triangles,
$$cg : EG :: ed : ED$$
or by substitution of equal values for these terms,
$$o'g' : O'G' :: r'd' : R'D'$$
and as $R'D' = R'E'$, the last proportion expresses the property of the ellipse that the ordinate of the ellipse is to the corresponding ordinate of the circle, as the semi-conjugate axis is to the semi-transverse axis.

The curve $cd—c'e'd'$ is therefore an *ellipse*.

THEOREM XII.—*The intersection of an oblique helicoid by a plane perpendicular to its axis, is a spiral of Archimedes.*

In Pl. XV., Fig. 115, revolve all the elements except $oO—o'o''$ about $O—O'O''$ until their horizontal projections coincide with $oO$. They will then all be parallel to V, and, as they make a constant angle with the axis, these revolved positions will be parallel to each other, in space, and hence in vertical projection also. Then, as the intersections of their vertical projections with $O'O''$ are equidistant, those with the ground line will be so also, making the distances of the latter points from $O'$ in arithmetical progression. Hence returning to their original positions, the distances of o, a, b, c, etc., from $O$ are in arithmetical progression. But the circular distances o1, o2, o3, etc., are in like progression. Hence the base oabc . . . of the helicoid, whose plane **H** is perpendicular to the axis, is a *spiral of Archimedes;* for the definition of this curve is that its successive radial distances $Oo$, $Oa$, etc., from a fixed point $O$, and its corresponding angular distances o1, o2, etc., from a fixed line $Oo$ are in arithmetical progression.

Continuing the process on the opposite side of $Oo$, the spiral terminates at $O$, tangent to that element which pierces **H** at $OO'$.

155. The last theorem affords a construction of the plane problem: *To draw a tangent to a spiral of Archimedes at a given point.* Let $dd'$ be the point. Construct an arc of the helix whose horizontal projection is the circle of radius $Od$ and which lies on a helicoid formed by drawing lines by (Theor. XII.) from o, a, b, c, etc., to meet $O$—$O'O''$ at any constant angle. Then the horizontal trace of a tangent plane at $dd'$ to this helicoid will be the tangent to the spiral at $dd'$.

EXAMPLE.—Make the construction here indicated.

156. Having a warped surface $ABEF$, Pl. XV., Fig. 118, if any tangent (raccording) hyperbolic paraboloid, as $BFkg$, be revolved 90° about its element of contact $BF$, it will become a *normal hyperbolic paraboloid* to the surface along the same element. That one of the tangent hyperbolic paraboloids, one of whose plane directors is perpendicular to $BF$, is a right one (128), **H**, to which $BF$ is parallel, being its other plane director. By revolving such a tangent parabo-

loid 90°, it becomes that one of the normal paraboloids, also a right one, whose *elements*, all perpendicular to *BF*, are *normals to the given surface*.

In other normal hyperbolic paraboloids at *BF*, it would only be tangents at points on *BF* to sections of them, made by planes perpendicular to *BF*, that would be normals to the given surface at those points.

157. If from a point in space tangents be drawn to the curves cut from a warped surface by a series of planes containing the given point, they will be elements of a *cone*, whose vertex is the given point, and which is tangent to the warped surface on the curve connecting the points of contact of the tangents.

158. If a warped surface be intersected by a series of planes, each parallel to a given line, tangents to the resulting curves, drawn parallel to that line will form a *cylinder*, tangent to the given surface and parallel to the given line, and whose curve of contact with the warped surface will contain the points of contact of the parallel tangents.

159. Any plane tangent to the cone (157) or to the cylinder (158) will be tangent to the warped surface to which these are tangent. This principle in connection with (76) indicates the solution of the problem: to draw a plane through a given line and tangent to a warped surface.

# PART II.—DOUBLE-CURVED SURFACES OF TRANSPOSITION.

160. *Double-curved surfaces of transposition* are of innumerable varieties. Among them we will only mention the *ellipsoid of three unequal axes*, Pl. XVI., Fig. 124, the *elliptic paraboloid*, the *elliptic hyperboloid* of separate nappes, and the *serpentine*.

The elliptic paraboloids and hyperboloids have elliptical sections perpendicular to the principal axis, in place of circular ones as in the analogous surfaces of revolution.

The *serpentine* is a double-curved helicoid, being generated by a sphere whose centre moves on a helix. That is, the surface of the serpentine encloses tangentially all the positions of the moving sphere. Any surface thus formed is called the *envelope* of all the positions of the generating surface. A cone of revolution, for example, is the envelope of all its tangent planes—that is, of all the positions of a plane which revolves about an axis with which it makes a constant angle at a fixed point.

### A—Projections.

PROBLEM LXXXIX.—*To construct the projections of a serpentine.*

*In Space.*—Imagine every point of a helix to be the centre of a position of the generating sphere. The two projections of each position will be equal circles. Curves tangent to the projections of all the spheres will be the projections of the serpentine.

*In Projection.*—Pl. XVI., Fig. 130. Let $beh$—$a'b'e'k'$ be the helix on which moves the sphere whose radius is $hm$—$h'm'$. The equal circles, shown only in vertical projection, represent positions of this sphere, their centres being at the points $aa'$, $bb'$, $cc'$, etc., of the helix.

The horizontal projection of the serpentine consists then of the concentric circles $Oo$ and $Om$, equidistant from $Oh$, that of the helix, by the distance $hm = ho$, the radius of the sphere. Its vertical projection consists of the curves tangent to the circles at $a'$, $b'$, $c'$, etc.

The curve, ascending from $l'$, suddenly stops at $n'$, descends to $r'$, and again ascends towards $O'$, in order to be continuous and yet tangent to all the circles. The like takes place at $u'$ and $s'$ and illustrates the occurrence, sometimes, of a *cusp* (99) in the projections of the complete contour of a surface.

The horizontal projection of the point $l'$ of the apparent contour is $l$, on $fl$, a parallel to $GL$. Other points of the horizontal projection of the apparent contour, seen in facing **V**, may be similarly found.

### B—Tangencies.

PROBLEM XC.—*To construct a plane through a given line and tangent to an ellipsoid of three unequal axes.*

*In Space.*—If any two points of the given line be made the vertices of two tangent cones, each circumscribing the ellipsoid, each cone will, by the known properties of the surface, have an ellipse of contact with the ellipsoid. Every plane, tangent to either cone, will also be tangent to the ellipsoid on its curve of contact with that cone. The two points of intersection of the two ellipses of contact will, therefore, be the points of contact of the two planes, each of which will be tangent to both the cones, and hence will contain the given line.

*In Projection.*—Pl. XVI., Fig. 124. Let the ellipses whose centres are $O$ and $O'$, and whose longer axes, only, are equal, be the projections of the given ellipsoid, $OC$—$O'$ being its intermediate, and $O$—$O'D'$ its shortest semi-axis; and let $Vv$—$V''v'$ be the given line. For convenience in making the planes of the ellipses of contact perpendicular to the planes of projection, take the vertices of the auxiliary tangent cones at $VV'$ and $vv'$, where the given line is cut by planes through $OO'$ and parallel respectively to **V** and **H**. Then $vcd$ is the horizontal projection of one of these cones, and $V''a'b'$ is the vertical projection of the other one, the omitted projections being unnecessary. Also $cd$ is the horizontal projection of the curve of contact of the former cone, and $cRS'$ is its plane; and $a'b'$ is the vertical projection of the curve of contact of the latter cone, and $b'NM$ is its plane.

The intersection, $cR$—$b'N'$ (Prob. X.) of these two planes intersects the surface of the ellipsoid, at its two points of intersection with either of the curves of contact.

In applying this solution to the sphere, these points are easily found, first in revolved position, the curves of contact being circles, but here it is equally convenient to find both projections of one of the curves of contact.

Accordingly, projecting $c$ and $d$ at $c'$ and $d'$, and $e$ at $e'$ and $e''$, we have four points of the vertical projection of the ellipse $cd$. Then $ff''$, projected from $f$ the middle of $cd$, bisects $c'd'$ and is an axis of this projection, limited by making $o'f' = o'f''$, a fourth proportional to $Oy$, $O'D'$, $fd$ ($Oy$, parallel to $cd$).

Having now its axes, the ellipse $c'f''d'f'$ can be easily found, and $p'p$ and $q'q$, its intersections with the ellipse $a'b'$, are then the points of contact of the required planes.

Finally, taking the point $pp'$, $Vp$—$V'p'$ is the element of contact of the tangent plane at that point with cone $VV'$, and $vp$—$v'p'$ its contact with cone $vv'$. Hence the traces, $k$ of the former, and $r'$ of the latter, and also the traces $h$ and

$g'$ of the given line, are points in the traces $kh$ and $r'g'$ of the tangent plane, $PQP'$, at $pp'$.

The tangent plane at $qq'$ may be similarly found.

161. The above solution is here applied to the general ellipsoid, instead of to the sphere, better to illustrate its utility. The method of Prob. LIII. would here have required the construction of tangents to an ellipse; while in that of Prob. LII. the plane of the *ellipse* of contact of a cylinder parallel to the given line would no longer have been perpendicular to the given line.

EXAMPLES.—1°. Construct the tangent plane at $qq'$.

Ex. 2°. Operate so as to find $p$ and $q$ before $p'$ and $q'$.

Ex. 3°. Apply the method of this problem to a sphere whose centre is in the ground line.

162. The curve of contact of the serpentine, Pl. XVI., Fig. 130, with one of its internally tangent spheres, will, in general, not differ sensibly from that great circle of the sphere which is perpendicular to the tangent to the helical directrix at the centre of the sphere.

The tangent plane to the serpentine at any point of this great circle will then be determined by the tangent to that circle at that point, and the tangent to the helix containing the same point.

### C—Intersections.

163. The ellipsoid of three unequal axes (Pl. XVI., Fig. 124) has two sets of circular sections. Those of each set lie in parallel planes, perpendicular to that of the longest and shortest axes, the central one of these planes for each set being determined by the intermediate axis, and a diameter equal to it.

Thus, the planes containing the axis $2OC-O'$ and the diameters $2O'x'$, or $2O'x''$, each cut the ellipsoid in circles; and by the known properties of the surface, all planes parallel to these will also cut the ellipsoid in circles. The centres of all the circles of each set are in the diameter which is conjugate (Theor. III.) to that, as $2O'x'$, or $2O'x''$, which determines the positions of the planes of these circles.

164. Let two cylinders of revolution, $C$ and $C_1$, have a common element of contact, $E$, and let one of them, $C_1$, be fixed, while the other, $C$, rolls upon it. The original element of contact, $E$, will then describe a species of surface of transposition, which will be a cylinder, being composed of the various parallel positions of $E$.

The right section of this cylinder is the curve generated by the point of contact, $e$, of the right sections, $c$ and $c_1$, of $C$ and $C_1$, as $c$ rolls on $c_1$. This curve is called an epicycloid, and if $c_1$ be straight, $C_1$ then being a plane, the curve is a cycloid.

Pl. XVI., Fig. 129, shows the epicycloid, $oabcd$, generated by the point of contact, o, of circles $O$ and $Ao4$, as circle $O$ rolls to the right from o to 4. The construction directly embodies the definition. Thus set off from o equal spaces on both circles; draw arcs with $Q$, centre of the fixed circle as a centre, through the points on circle $O$, and through $O$; note $O_1$, $O_2$, etc., the intersection of radii through $Q$ and the points 1, 2, 3, 4, on the fixed circle, with the arc $OO_4$; draw circles, equal to circle $O$, with $O_1$, $O_2$, etc., as centres, and their intersections with the arcs through the corresponding points, 1, 2, 3, on circle $O$, will be the points, $a$, $b$, $c$, etc., of the epicycloid.

165. Let $C$ and $C_1$ be two cones of revolution, having a common vertex, $V$, and element of contact, $E$, and let $B$ and $B_1$ be the bases, or right sections, which are tangent to each other at any point $p$, on $E$.

This supposed, let $C$ roll upon $C_1$, which remains fixed. The point $p$, remaining a point of $B$, rolling on $B_1$, will generate a kind of epicycloid, which, being everywhere equidistant from $V$, lies on the surface of a sphere whose centre is $V$, and radius $Vp$. The curve is, therefore, called a *spherical epicycloid*. It is the intersection of the derivative or secondary cone, $C_n$, generated by the element $E$, with the sphere of radius $Vp$.

PROBLEM XCI.—*To construct the projections of a spherical epicycloid, and of a tangent line to it at any point.*

*In Space.*—The generatrix $p$ (165) of the curve is at any moment in the circumference of the base $B$ of the rolling cone $C$, and also in the common meridian plane of the two cones through the common element of contact at the same moment, and its position is constructed according to these principles.

*In Projection.*—1°. Pl. XVI., Fig. 128. Let $VV'$ be the common vertex of the fixed cone $Vfm - VaV'$ and the rolling cone $V'aK'$, and let $f$ be the original point of contact of their bases.

Revolve the base $aK'$ about the common tangent at $a$, into **H**, at $af_1K''$, and then lay off from $a$, on both circles, equal spaces $ab$, $a_1b_1$, etc., one of which shall exactly divide $af$. Then $f_1$ is the point that was originally at the point of contact $f$; hence project $f_1$ at $f_1'$, and counter-revolve it to $FF'$, which is the point of the epicycloid reached by $f$ when $a$ is the point of contact of the cones' bases.

2°. *To find other points*, we may proceed in either of two ways:

*First.* To find the point of the epicycloid, corresponding, for instance, to $V'Vc$ as the common meridian plane of the two cones, take $V'Vc$ as a new vertical plane of projection, and then proceed with that plane just as with **V** in finding the point $FF'$.

*Second.* Imagine $V'Vc$ revolved about $V'V$ till it coincides with **V**, find $D_1D_1'$, exactly as $FF'$ was found, since $d_1$ is the third point from $a_1$ as $f$ and $f_1$ are from $c$ and $c_1$, and then counter-revolve $V'Va$ back to its original position, $V'Vc$, when $D_1D_1'$ revolving through an equal angle, will appear as at *DD'*, the point reached by $f$ ($f_i$) when $c$ and $c_i$ are in contact.

3°. *The tangent line*, at any point of the epicycloid, will be the intersection of the tangent plane at the given point to the sphere containing the epicycloid, with the tangent plane to the sphere whose centre is the corresponding point of contact of the cones' bases, and whose radius is the line from that point to the given point.

The latter sphere is properly chosen, since the generatrix is at each instant evidently describing a momentary arc, for that instant only, about the point of contact of the cones' bases for the same instant.

EXAMPLES.—1°. Construct the tangent line just described.

Ex. 2°. Find a point of the epicycloid corresponding to any position as *V'Vm* of the common meridian plane of contact.

Ex. 3°. Construct the tangent at a given point on the epicycloid, Pl. XVI., Fig. 129. [The tangent at *c*, for example, is perpendicular to the line *cr*.]

PROBLEM XCII.—*To find the intersection of a cone and sphere, the centre of the sphere being at the vertex of the cone, together with a tangent to the curve.*

*In Space.*—The auxiliary planes employed in the solution may either be perpendicular to a plane of projection and through the cone's vertex, or parallel to a plane of projection. In the former case, they will cut elements from the cone, and great circles from the sphere. In the latter case, they will cut sections similar to the base from the cone, and small circles from the sphere. In each case, the line cut from the cone will intersect that cut from the sphere by the same plane, in points common to both surfaces—that is, in points of their intersection.

*In Projection.*—Pl. XVI., Fig. 125. Let *VFBN—V'A'B'* be the cone, oblique with a circular base. The arc *S'S"* with *V'* for its centre sufficiently represents the sphere.

1°. *The highest and lowest points.*—These are on the longest and shortest elements, *VH* and *VK*, in the vertical plane

$VH$ containing the axis $VO$—$V'O'$. Revolving the plane $VH$ about a vertical axis at $V$ till parallel to **V**, the element $VH$—$V'H'$ will appear at $VH''$—$V'H'''$. The similarly revolved vertical great circle of the sphere contained in the plane $VH$ will intersect $VH''$—$V'H'''$ at $b'''$, the revolved position of the highest point. In counter-revolution, this point returns by the horizontal arc vertically projected in $b'''b'$ to $b'$ on $V'H'$, whence it is projected down to $b$ on $VH$. Otherwise, $b'''$ might have been projected on $VH''$ and the counter-revolution shown first in horizontal projection by an arc with $V$ as a centre, giving $b$ first.

The lowest point may be similarly found.

2°. *The points on the right and left elements.*—These elements are horizontally projected in $VA$ and $VB$ ($AB$ parallel to $GL$). The points on them are found as just explained, and as shown at $aa'a'''$ the point on $VA$—$V'A'$.

3°. *The points on elements parallel to* **V**.—These elements are $VC$—$V'C'$ and $VD$—$V'D'$, and the points on them, $cc'$ and $dd'$, are immediately found without revolution, $S'S''$ being itself that great circle which is in their plane.

4°. *The points on the extreme elements seen on* **H**.—These elements are horizontally projected in $VE$ and $VF$, and the points on them (not shown) are found by revolving them till parallel to **V**, as before.

5°. *Points in horizontal auxiliary planes.*—$m'v'$ is such a plane. It cuts from the cone a circle of centre $oo'$, on the axis $VO$—$V'O'$, and radius $o'q'$; and from the sphere a circle of radius $m's'$—$Vs$. The horizontal projections of these circles intersect at $r$ and $n$, whose vertical projections $r'$ and $n'$ are on $m'v'$.

This method obviously applies only to points between the highest and lowest. At the latter, the circles would be tangent to each other, but the assumed plane could only be accidentally at the proper height to contain them.

6°. *The tangent to the intersection.*—This, at any point, as $nn'$, is the intersection of two planes, one tangent to the sphere at that point, the other to the cone, along the element through the same point. The former plane is found as in (Prob. LI.), the latter as in (Prob. XXVIII.).

Thus $V,x$, the intersection of a tangent (not drawn) to $S'S''$ at $ss'$, with the line $VV'$, is the vertex of a cone tangent to the sphere on the horizontal circle $m's'$. Then $xn'—Vn$ is the element of this cone containing $nn'$, and $t't$ is its horizontal trace. Hence $Tt$, perpendicular to $Vn$, is the horizontal trace of the tangent plane to the sphere at $nn'$. But $NT$ is the like trace of the tangent plane to the given cone on the element $VnN—V'n'$, hence $Tn$, the horizontal projection of the intersection of these two planes, is the horizontal, and $Tn'$ is the vertical projection of the required tangent.

EXAMPLES.—1°. Find all the points described but not shown on the figure.
Ex. 2°. Find the points in an auxiliary plane perpendicular to V.

## D—Development.

PROBLEM XCIII.—*To develop the surface of an oblique cone.*

*In Space.*—All points of the intersection of the cone and sphere, Prob. XCII., being on the surface of a sphere, are equidistant from its centre; this curve will therefore, in the development of the cone, appear as a circular arc with a radius equal to that of the sphere. It only remains to determine the length of this arc as a means of *locating* the developed elements. The *lengths* of these can then be found as in (Prob. XII.).

To find the length of the intersection of the cone and sphere, we develop either of its projecting cylinders.

*In Projection.—First.* Pl. XVI., Fig. 126, represents the development of the cylinder whose base is the horizontal projection, *nbcrda*, Fig. 125, of the intersection of the cone and sphere, and whose elements, $nn'$, $bb'$, etc., are the heights, $n'n''$, $b'b''$, etc., of the points of the curve, above **H**. Then *nbcrda*, Fig. 125, develops into the straight line $nn_{,,}$

Fig. 126, and $n'b'a'n_1'$ is the true length of the intersection itself.

*Second.* Fig. 127. With $V$ as a centre and $V''S'$, Fig. 125, as a radius, describe the arc $n'n_1'$, and make it, and its divisions, equal to $n'n_1'$ and its divisions in Fig. 126, and in like order of parts. Then $n'n_1'$, Fig. 127, is the development of the intersection of the cone and sphere, as seen on the developed cone. Lines from $V$ through $n'$, $b'$, etc., will then be developed elements, where $VH = V''H'''$, Fig. 125, likewise $VA = V''A''''$, and $VC$ and $VD$ respectively equal $V''C'$ and $V''D'$ in Fig. 125, etc., giving $VNHDN'$ for the required development of the cone in Fig. 125.

To develop the tangents $Tn$ and $TN$, Fig. 125, draw $n'T$ perpendicular to $VN$, and limit it by an arc from $N$ as a centre with radius $NT = NT$, Fig. 125.

# SHADES AND SHADOWS.

### First Principles.

166. An opaque body, $B$, Pl. XVII., Fig. 1, being exposed to light which proceeds in any direction, as $Rr$, will be partly illuminated and partly dark.

The unilluminated portion $CdT$ is called the *shade* of the body, and the boundary $CPTO$ between this portion and that which receives light, is the *line or curve of shade*.

167. The portion of space from which light is excluded by the opaque body just supposed, is the *shadow in space* of that body.

When any other surface $Q$ intercepts this shadow, the portion of it, $cpto$, which is thus deprived of light by the opaque body, is the shadow of that body on that surface, and the boundary, which determines this shadow, is called the *line or curve of shadow*.

168. *Any point, $P$, of the curve of shade* of a body is the point of contact of a ray of light tangent at that point to the body, for any other ray, not tangent to the body, would either intersect it as $Mn$, at $n$, in a point therefore illuminated, or would be wholly exterior to it as at $Rr$.

169. Neither of the two rays just mentioned could give any point of the curve of shadow of the opaque body; the exterior ray, because it nowhere touches the body, nor the secant ray, because it would be stopped by the body.

Hence any point of the shadow of a body $B$, upon another surface $Q$, is the intersection of that surface with a ray tangent to that body. That is, the *curve of shadow*, as *cpto*, of a body is the *shadow of its curve of shade*, as *CPTO*.

Hence in the construction of shades and shadows, the shades (166) must be found first.

170. Light proceeds from a luminous point in every direction, in straight lines called *rays*.

The surface of rays radiating from such a point, and enclosing an opaque body tangentially, will be a *cone*, when the luminous point is at a finite distance from the body, and a cylinder, $CcTt$, Fig. 1, when the rays are parallel, as they are when proceeding from a body at a vast distance, as the sun, to the points of any terrestrial object.

171. The character of Shades and Shadows as an application of Descriptive Geometry is now apparent.

The *curve of shade of any body is the curve of contact of that body with a circumscribed tangent cylinder or cone*, according respectively as the source of light is very remote or quite near.

*The curve of shadow* of a body is the curve of intersection of the cylinder or cone just mentioned with whatever surface receives the shadow.

The terms cylinder and cone are here used in the widest sense as meaning respectively surfaces composed of parallel or of converging elements. In the case of any plane-sided opaque body, as a cube, the tangent cylinder and cone of rays would become respectively an enclosing prism and pyramid of rays, and the line of shade would consist of the sum of those edges of the body which separated its illuminated from its unilluminated faces.

172. The manner of finding either the *contact* or the *intersection* of a straight line, or ray, element of the tangent surface of rays, with a given surface, depends largely on the nature of that surface. Hence *problems of shades and shadows are naturally classified like those of descriptive geometry;* giving shades and shadows on plane, developable, warped, double-curved surfaces.

173. Finally, in order that the reality indicated in Pl. XVII., Fig. 1, may be exactly shown on paper, the *source of light*, the *body casting* the shadow, and the *surface receiving* it, must all be given by their projections. The source of light, if near, will be given by its projections; if remote, by the two projections of any ray, to which all the others will be parallel.

174. In making drawings for practical purposes, the light is usually taken in one fixed direction viz., so that its *projections* make angles of $45°$ with the ground line, as at $A_1a'$—$A'a'$, Pl. XVII., Fig. 2.

The corresponding direction of the light itself is evidently that of the diagonal of a cube, one of whose faces coincides with **H**, and the other with **V**, the diagonals of those faces being the projections of the ray, and all converging towards the right to one point on the ground line. The ray itself thus makes an angle of $35° 16'$ with **H** and with **V**.

In general problems, to illustrate principles, the light is taken in any direction that is most convenient.

## A—Shades and Shadows on Planes.

PROBLEM I.—*To find the shades and shadows on a rectangular block having a panel in front and a tablet on top; all its edges being parallel or perpendicular to* **H** *and* **V**.

*In Space.*—The edges of *shade* (166) in all such cases, and in many others, can be determined by inspection.

The shadow of such a body consists of the shadows of its limiting points and edges.

The shadow of a point upon a plane is where the ray through that point pierces that plane.

The shadow of a straight line upon a plane is the trace upon that plane of a plane of rays containing the line.

*In Projection.*—Let Pl. XVII., Fig. 2, represent the block with its panel and tablet.

As a temporary aid to the learner while becoming familiar with shadow construction, we will, in a few of the earlier problems, denote points and their shadows by the same letter—the former by a capital, the latter by a small letter.

$Aa$—$A'a'$ taken at 45° (174) being the projections of a ray of light, the edges of shade (166) are evidently the vertical ones at $A$ and $C$, and the horizontal ones, $AB$—$A'$ and $BC$—$A'C'$ of the top of the block; together with the corresponding edges of the tablet, with $H$—$K'H'$ and $HI$—$H'I'$ of the panel.

1°. *To find the shadow of a point upon* **H**.—Let $AA'$, the front, right, upper corner, be the point. Through this point draw the ray $Aa$—$A'a'$, which (D. G., Prob. VIII.) pierces **H** at $a$, which is therefore the shadow of $AA'$.

2°. *To find the shadow of a point upon* **V**.—Let $B, A'$ be the point through which draw the ray $Bb$—$A'b'$, which pierces **V** at $b'$, which is therefore the shadow of $B, A'$ on **V**.

3°. *To find the shadow of a vertical line on* **H**.—Let $A$—$D'A'$ be the line. The point $AD'$ being in **H**, is its own shadow at $A$, and $a$ is the shadow (1°) of $AA'$. Hence $Aa$ is the shadow of $A$—$D'A'$. Likewise $Cc$ is the shadow of a part of $C$—$R'C'$ on **H**.

4°. *To find the shadow on* **V** *of a line perpendicular to it.*—Let the portion $BA_1$—$A'$ of $AB$ be the line, having drawn the ray $a'A_1$ which determines the point $A_1$ of which $a'$ is the shadow. Then $b'$ being the shadow of $B, A'$ on **V**, $a'b'$ is the shadow of $A_1B$—$A'$ on **V**.

5°. *To find the shadow on* **H** *of a line parallel to it.*—Let $AB$—$A'$ be the line. Then $a$ being the shadow of $AA'$ (1°), and $b_1$ that of $B,A'$ on **H**, supposing **V** for the moment to be transparent, $ab_1$, obviously equal and parallel to $AB$, is the shadow of $AB—A'$ on **H**.

6°. *To find the shadow on **V** of a line parallel to it.*—Let $C—R'C'$ be the line. Then (2°) $c'$ being the shadow of $CC'$ on **V**, and $r'$ that of $C,R'$ on the lower part of **V**, supposing **H** transparent, $c'r'$, evidently equal and parallel to $C—R'C'$, is the shadow on **V**.

Likewise $b'c'$ is the shadow of $BC—A'C'$ on **V**.

The shadows, $Ee$ of $E—M'E'$; $ef$ of $EF—E'$; $fg$ of $FG—E'G'$, and $gG$ of $G—G'N'$, all on the top of the block; likewise $h'l'$ of a part of $H'K'$, and $h'j'$ of a part of $H'I'$ on the panel, being found in the same manner as the preceding, need no further explanation.

If we revolve the vertical plane of rays on $Aa$ about its horizontal trace $Aa$ into **H**, the line $A—D'A'$ will appear as a perpendicular $AA''$ (not shown) to $Aa$, at $A$, and equal to $A'D'$; $A''a$ will then show the true length of the ray $Aa—A'a'$, and $AaA''$ will be the angle made by the ray with **H**.

EXAMPLES.—1°. Let the block be so far from **V** that its shadow will be wholly on **H**.

Ex. 2°. Let it be so far above **H** that its shadow will be wholly on **V**.

Ex. 3°. Let the tablet be on the front, and the panel in the top.

Ex. 4°. Find the angle made by the light with **V**.

Ex. 5°. Let the tablet on top be high enough to cast its shadow partly on **H** and **V**.

Ex. 6°. Let an opening of the size of the panel extend through the block from front to back.

Ex. 7°. Let there be tablets on the two ends of the block.

175. *Useful elementary principles* derived from the last problem.

1°. *The shadow of a vertical line* on **H** is in the *direction of the horizontal projection of the light;* for it is the horizontal trace of the vertical plane of rays through that line.

2°. Likewise the shadow on **V** of a perpendicular to it, is in the direction of *the vertical projection* of the light.

3°. The shadow of a line on a plane parallel to it is equal and parallel to the line.

4°. Parallel lines will have parallel shadows on the same, or on parallel planes. Also the shadows of the same line on parallel planes are parallel.

5°. Whenever the *direction* of a shadow is known, *one point* of it will locate it. Otherwise two will be necessary.

6°. When the shadow of a line falls partly on each of two intersecting planes, the point on the intersection is common to both shadows (see $a'$, Fig. 2), for the two traces ($aa'$ and $A'a'$ of the plane of rays through the line $AB$—$A'$) necessarily meet that intersection at the same point.

PROBLEM II.—*To find the shades and shadows of a stone cross.*

*In Space.*—The edges of shade (166) can still be found mostly by inspection, or certainly by means of *planes of rays through the vertical edges.*

Some of the edges of the cross being oblique to the planes which receive their shadows, we must apply (175, 5°).

*In Projection.*—Pl. XVII., Fig. 3. 1°. *The Shades.*—The visible edges of shade are those that are made heavy. Under the theory, however, that these edges of shade are, strictly, the elements of contact of planes of rays with the minute quarter cylinders, which, instead of sharp mathematical lines, really form the edges of the body, such elements, when situated as at $G$, would be invisible in facing **V**. The vertical edge at $G$, and other similar ones, are therefore inked lightly.

The vertical edges at $D$ and $G$, for example, are certainly known to be lines of shade by means of the vertical planes of rays, $Dd$ and $Gg$, containing them, since these show the vertical faces on $KD$ and $KG$ to be in the light, and those on $FG$ and $FD$ to be in the dark.

2°. *The Shadows.*—Beginning with $AA'$, its shadow,

$Aa$—$A'a'$ being the direction of the rays, is $a$ (Prob. I., 1°). That of $A,B'$ is $b$, similarly found, whence $ab$ is the shadow, on **H**, of $A$—$A'B'$ (175, 1°). Likewise that of $CC'$ is $c$, also found by making $bc$, from $b$, already found, equal and parallel to $AC$ (175, 3°). Thence $ch$ equal and parallel to $CH$ is the shadow of $CH$—$C'H'$, interrupted at $i$ and $j$ by the shadows, $Di$ and $Gj$, of the lower parts of the vertical edges at $D$ and $G$. $Gl$ is the shadow of a part of $G$—$G'N'$ on the top, $HF$—$H''N'$, of the horizontal arm. $hh''$ is the shadow on **H** of a part of the vertical edge, $H$—$H'H''$. The last four shadows, with $de$, that of the upper portion of the vertical edge $D$—$D'E'$, are all parallel to $ab$ (175, 1°, 4°).

Returning to $a$, the shadow $ad$, of $AD$—$A'D'$, on **H**, is equal and parallel to that line.

Entering **V**, we find $e'$ the shadow of $EE'$ (Prob I., 2°), whence $e'e$, parallel to $D'E'$, is the shadow on **V** of the upper part of that line. Then $f'$, being the shadow of $FF'$, gives $e'f'$ that of $DF$—$E'F'$; and $f'g'$, similarly found, as shown, is that of $FG$—$F'G'$. The remaining points of shadow on **V**, which can be similarly found, are invisible.

Finally, the portion $KO$—$K'O'$ casts a shadow on the vertical face on $KD$, beginning at $KK'$, which is its own shadow on that surface, and ending at $D,o'$, the shadow of $OO'$ on the vertical edge at $D$.

EXAMPLES.—1°. Turn the cross horizontally till the face at $K'D'$ is in the dark.
Ex. 2°. Turn it till $HC$ is parallel to **V**.
Ex. 3°. Keeping it as it is, change the direction of the light.

**PROBLEM III.**—*To find the shadow of a turnstile on* **H** *and on a vertical plane oblique to* **V**.

*In Space.*—The new given plane being vertical, its horizontal trace may be considered as the ground line, and the operations will be precisely similar to those of the last two problems.

*In Projection.*—Pl. XVII., Fig. 4. The construction being fully shown, the explanation is left to the student.

EXAMPLES.—1°. Removing the fence, find the shadow of the turnstile wholly on **H**.

Ex. 2°. Placing the turnstile near **V**, find its shadow on **H** and **V**.

Ex. 3°. Draw the figure inverted, but with $AB$ and the rays *then* inclined to the right of $A$, $C$, etc., which will suit the enunciation: To find the shadow of the axis, perpendicular to **V**, and arms, in a vertical plane, upon a roof slope $AB-A'B'$ perpendicular to **V**.

PROBLEM IV.—*To find the shadow cast on a flight of steps by one of its piers.*

*In Space.*—The new features in this problem are—*first*, lines of shade oblique to both planes of projection; *second*, narrowly limited surfaces receiving the shadow; *third*, an auxiliary elevation on a plane perpendicular to the ground line.

The *second* feature will serve to show the convenience in such cases of finding points of shadow in surfaces produced, but that the real portion of the shadow is only on the real portion of the surface. The *third* will illustrate that, in general, *any two* of three projections will be sufficient for the purpose of finding the shadows on all.

*In Projection.*—1°. *Preliminary.*—Pl. XVII., Fig. 5. This is a flight of three steps, distinguished by numbering their tops 1, 2, 3. The pier at the left has a level top, $CDK-C'K'-C''D''$, a slope, $AC-A'CK'-A''C''$, and a front, $AB-H'B'A'-A''B''$, also sloping. The side elevation, $B''D''$, is on the plane $TUT'$, where the hidden profile of the steps is dotted. Thus $A'$ and $A''$ are on the same horizontal line. Also like points on the side elevation and the plan are equidistant from the trace $T'U$, and from the ground line $H'U$, respectively. Thus $B'B = UB''$, etc.

The edges of shade of the pier which cast shadows on the steps, and on the ground in front of step 1, are $AB-A'B'-A''B''$, $AC-A'C'-A''C''$, $CD-C'-C''D''$.

The light is taken in the conventional direction of 45° in *projection.* Thus $Aa$—$A'a'$—$A''a''$ are the three projections (174) of one ray, to which all others are parallel.

2°. *The shadows of $AB$—$A'B'$.*—Beginning at $BB'$, this point being in **H**, is its own shadow on **H**. The shadow of $AA'$ on **H** is $aa'a''$, where the ray $Aa$—$A'a'$—$A''a''$ pierces **H**. Then $Bn$, so much of $Ba$ as is in front of step 1, is the shadow cast by a part of $AB$—$A'B'$ on the ground—that is, on **H**.

The shadow of $AA'A''$ on the front of step 1 is $a_1a_1'a_1''$. Here (*In Space, Third*) $a_1'$ can be found, either by projecting up $a_1$ upon $A'a'$, or by projecting over $a_1''$ upon $A'a'$. Then, projecting $n$ at $n'$, gives $n'a_1'$ the shadow of a part of $AB$—$A'B'$ on the front of step 1.

3°. *The remaining shadows on step 1.*—The ray $Aa$—$A'a'$ pierces the plane $A'c_1'$ of the top of step 1 at $a_2'$, which, projected down on $Aa$, gives $a_2$. The ray $Cc$—$C'c'$ pierces the same plane at $c_1'c_1$, hence $a_2c_1$ is the indefinite shadow of $AC$—$A'C'$ on this plane, and $ep$—$e'p'$, so much of $a_2c_1$ as is within the limits of the top of step 1, is its real portion. Then by (175, 6°) $ea_1$—$e'a_1'$—$e''a_1''$ is the remainder of the shadow on the front of step 1.

The point $ee'$ can, however, be found directly by beginning with the projection $e''E''$ of the ray which meets the front edge of step 1. Then project $E''$ over to $E'$ on $A'C'$, and draw the projection $E'e'$ of the same ray, which gives $e'$.

4°. *The remaining shadows.*—$cs$ projected down from $c'$, and parallel to $CD$, is the shadow of $CD$—$C'$ on the top of step 3. $qr$ and $vc$ are parallel to $ep$, and by reason of the equality of the steps, $c$, $q$ and $v$ are all in the same line, parallel to $AC$—$A'C'$—$A''$,$C''$, since this line is parallel to the plane $q''e''$ of the front edges of the steps. Likewise, $pq$—$p'q'$ and $rv$—$r'v'$ are parallel to $a_1e$—$a_1'e'$.

5°. *Plan construction from the side elevation only.*—Take, for example, the point $r''r$, the point of shadow on the bottom edge of step 3 (the same thing as the back edge of step 2). Drawing the ray $r''R''$, we find the point, $R''$, whose shadow is $r''$, project $R''$ at $R''''$; by counter-revolution about $U$ as a centre, bring $R''''$ to $R'''$, whence project it by $R'''R$ parallel to the ground line to $R$, the horizontal projection of $R''$. Then the horizontal projection $Rr$ of the same ray, $R''r''$, intersects the given edge of step 2 at $r$, the horizontal projection of $r''$. Similarly, $a$ is found from $a''$, $c$ from $c''$, etc.

EXAMPLES.—1°. Change the proportions of the steps, also enlarge the scale.

Ex. 2°. Change the direction of the light (174).

Ex. 3°. Let the piers of the steps converge towards the back of the steps—that is, let the plane $BB'C$, continuing vertical, be oblique to **V**.

176. Summing the four last problems, we deduce the following principle and consequent rule for the solution of every problem of *shadows upon the planes of projection, or upon any plane parallel or perpendicular to them.*

*One projection of every such plane is a line,* viz., that upon the plane of projection to which the given plane is perpendicular. Thus, $AB$, Fig. 4, is the horizontal projection of the face of the fence, $A'B'$; $e'c_1'$, Fig. 5, is the vertical projection of the top of step 1, etc.

Hence, comparing the similar constructions of $a$ and $b'$, Fig. 2; $a$ and $c'$, Fig. 3; $f'$, etc., Fig. 4; $a,c_1a_1'$, etc., Fig. 5, we have for all like cases the

RULE.—Note first the intersection of *that* projection of the given plane which is a *line*, with the *like* projection of the ray employed (see $c'$ or $a_1$, Fig. 5, etc.), and then project the point so found upon the other projection of the same ray. (See $c$, or $a_1'$, Fig. 5, etc.)

177. All the lines casting shadows have thus far been straight; but the operations would have been precisely similar had those lines been curved. That is, the shadow of any curve upon any plane situated as in (176) is found by assuming any convenient number of points upon it, and finding their shadows, just as in the preceding problems, and then joining them.

The shadow of any plane curve upon a plane parallel to it will, in general, be an equal curve. That of a circle, on any other than a parallel plane, is generally an ellipse. If, however, the plane of any curve be parallel to the direction of the light, it will be a plane of rays, and its trace, the shadow of the curve, will be straight.

EXAMPLES.—1°. In Pl. XVII., Fig. 2, let the tablet be replaced by a horizontal circle at any height from the block.

Ex. 2°. Also by any vertical semicircle having its diameter in the top of the block.

Ex. 3°. Let the panel be cylindrical.

Ex. 4°. In Fig. 3, let the arms bear semicircles on $DF$, $AC$, and $A'B'$, as diameters.

Ex. 5°. In Fig. 4, let the arms of the turnstile be curved.

Ex. 6°. In Fig. 5, let an arc, tangent to $D''C'$ and $B''A''$ at $C'$ and $A''$, be the profile of the pier.

178. It is only important to add that in constructing the shadows on different vertical projections (elevations) of the same fixed object, the light is supposed to turn equally with the observer as he turns from facing one vertical plane to view another. Thus, if, in Pl. XVII., Fig. 5, the side elevation bore any visible shadows, the three projections of the light used in finding them would be in the direction of the three arrows $r_1$, $r_1'$, $r_1''$, which indicates a revolution, by the light, of 90° around a vertical axis, as the observer also turns 90° from viewing the vertical plane **V**, to viewing $TUT'$, which is $\mathbf{V}_1$.

PROBLEM V.—*To find the shadows of chimneys on side and end plane roof surfaces.*

*In Space.*—Here most of the surfaces are not perpendicular to either plane of projection, and hence will not have any linear projection. Hence we find where a ray pierces any of them by (D. G., Prob. XI.), that is, by passing, not a ray alone, as in the four preceding problems, but a *plane of rays also*, through a point casting a shadow. Then, *where the ray through that point pierces the trace upon the roof of the plane of rays through the same point will be the shadow of that point upon the roof.*

*In Projection.*—Pl. XVII., Fig. 6. $ROMQNS$ is the horizontal, and $M'N'S'Q'R'$ the vertical projection of the roof, and $ABC$—$T'A'C'$ and $DF$—$o'D'F$ are the chimneys whose shadows upon the roof are to be found.

1°. *The shadow of the left-hand chimney.*—Taking the plane of the foot of the roof as **H**, and the edge $AB$—$A'$ perpendicular to **V**, and $Aa$—$A'a'$ for the direction of the light, $A'h'h$ is the plane of rays containing $AB$—$A'$, and therefore perpendicular to **V**. It cuts the ridge of the roof at $g'g$, and its foot at $h'h$ and $h'h''$, giving $hg$—$h'g'$ and $h''g$—$h'g'$ as its traces on the front and back roof slopes. Then by (D. G., Prob. XI.) the rays $Aa$—$A'a'$ and $Bb$—$A'b'$ meet the roof at their intersections with these traces, giving $aa'$ and $bb'$ as the shadows of $AA'$ and $BA'$, and thus $agb$—$a'g'b'$ as the shadow of $AB$—$A'$ upon the roof.

The edge $BC$ being parallel to the side roofs, its shadow, $bc$—$b'c'$, is equal and parallel to $BC$—$A'C'$. Finally, as the edges at $A$ and $C$ are vertical, the planes of rays containing them are so also, and consequently their shadows are horizontally projected (D. G., 27, 8°) in the horizontal traces $Aa$ and $Cc$ of these planes. The vertical projections of these shadows are $a'p'$ and $c'q'C''$.

2°. *The shadow of the right-hand chimney.*—$ODs$ and $FEt$ being the traces of the front and back of this chimney upon the roof, project $s$ and $t$ at $s'$ and $t'$, and draw $s'o'o''$ and $t'F''$ parallel to the ground line $M'R'$, to obtain the vertical projections of these traces.

Then $Dn$—$o'n'$, the shadow of a part of $D$—$o'D'$ on the front roof, is parallel to $Aa$—$p'a'$ (175, 4°), $dd'$ is the shadow of $DD'$ on the plane of the roof surface $SQR$—$S'Q'R'$ (Prob. IV.), and is found as $aa'$ was. That is, $D'm'm$ is the plane of rays containing $DE$—$D'$, and $m''l$—$m'l'$ (parallel to $h''g$—$h'g'$), and $lm$—$l'm'$, are its traces on the back and end roofs, intersected by the rays $Dd$—$D'd'$ and $Ee$—$D'e'$ at $dd'$ and $ee'$, the shadows of $DD'$ and $EE'$. Hence $nv$—$n'v'$, part of $nd$—$n'd'$, is the remainder of the shadow of $D$—$o'D'$.

Finding $ff'$ as $bb'$ was found, $Fjf$—$F''j'f'$ is the shadow of $F$—$F''F'$; thence $fk$—$f'k'$, parallel to $FE$—$F'D'$, is the shadow of a part of that line, and $ke$—$k'e'$ is the remainder of its shadow.

By drawing a ray inversely from $kk'$ to meet $FE$—$F'D'$, we shall find what part of this line has $fk$—$f'k'$ for its shadow.

All the shadows on the back part of the roof, or behind the right-hand chimney, are invisible in vertical projection.

EXAMPLES.—1°. Vary the proportions of the roof in any manner.

Ex. 2°. Change the direction of the light.

Ex. 3°. Let $NS$ and its parallels be oblique to V.

Ex. 4°. Add an abacus to each chimney [see Fig. 7, which indicates that the portions $ED$ and $AF$, only, of the *front and left-hand under edges* of the abacus, cast shadows on external objects, the portions $AB, BC$ casting shadows on the front, and $CD$ on the left side of the chimney].

Ex. 5°. A very interesting example is found by means of the following modifications of Pl. XVII., Fig. 6. Let each chimney be cylindrical, and have a cylindrical abacus. It will be convenient to make the figure large enough to fill half the plate or more, and to use an auxiliary vertical projection on a plane perpendicular to $M'R'$ and revolved into H.

179. Fig. 6 shows clearly how to ascertain which of a group of intersecting plane surfaces in various positions are in light or shade. For example, had the ridge been raised to the height $g_1'N_1'$, and the direction of the light changed to $Aa_1$—$A'a'$, the traces of the plane of rays $A'h'h$ on the new roof slopes would have been $g_1h$ and $g_1h''$, and the same ray $Aa_1$ would have pierced both of these traces, at $a_1$ and $a_2$. *In every such case, the surface first pierced by the ray is in the light, and the other is in the dark, and their intersection ($g_1'N_1'$) is an edge of shade or shade line* (166). Likewise, had the ray $Dd$—$D'd'$ pierced the front roof before piercing $QSR$, the latter would have been in the dark, and $QS$ an edge of shade.

On the other hand, with the figure as it is, the whole of

the roof is in the light, except as shadowed by the chimneys.

The principles and problems now given will enable the student to solve every possible problem of shadows on plane surfaces, whether cast by straight or curved lines.

## B—Shades and Shadows on Developable Surfaces.

180. The elements of shade on a cylinder and cone, the only developable surfaces necessary to consider here, are the elements of contact of tangent planes of rays.

When the luminous point is near, the problem therefore merely becomes an application of (D. G., Prob. XXX.). When it is indefinitely remote, and the rays consequently parallel (173), the problem is an application of (D. G., Probs. XXXI., XXXII.), the tangent plane in each case being then a plane of rays.

EXAMPLES.—1°. Given a cylinder oblique both to H and V, and the direction of the light, to find the elements of shade on its convex surface.

Ex. 2°. Find the elements of shade on an upright cone.

Ex. 3°. Find the elements of shade upon an inverted cone.

Ex. 4°. Find the elements of shade upon a cylinder in any horizontal position, perpendicular, parallel, or oblique to V.

Ex. 5°. Find the elements of shade upon a cone in the same position.

Ex. 6°. Find the elements of shade upon a cone whose axis is oblique both to H and V.

PROBLEM VI.—*To find the shadow of a rectangular abacus upon a cylindrical column and the shade of the column.*

*In Space.*—A cylindrical surface being the only one besides a plane which has parallel elements, it is the only curved surface capable of being perpendicular to a plane. When thus situated, its projection upon the plane to which it is perpendicular is a curved line, viz., its right section. The shadows of any points on a cylinder so situated may therefore be found in the same manner as upon planes similarly placed.

*In Projection.*—Let $EAD$—$A'D'$, Pl. XVIII., Fig. 10, be the square abacus of a cylindrical column, half shown at $FbG$—$FGG'$, $Cc$—$C'c'$ being the direction of the light; $Cc$ is also the horizontal trace of the vertical tangent plane of rays, whose element of contact $c$—$d'c'$ is the visible element of shade of the column. The portions $CA$ and $AE$ of under edges of the abacus then cast shadows on the column. Thus the ray $Aa$—$A'a'$ pierces the column at a point whose horizontal projection is $a$, and whose vertical projection, $a'$, is found by projecting $a$ upon $Aa'$, giving $aa'$ the shadow of $AA'$ upon the column. Any other points, as $bb'$, can be similarly found. $a'f'$ is the vertical projection of the intersection of the plane of rays containing $AE$—$A'$, and thence perpendicular to **V**, with the cylinder; that is, $aF$—$a'f'$ is the shadow of $AE$—$A'$ on the cylinder.

181. From the foregoing solution, we have at once the rule for finding shadows on all cylinders which are perpendicular to a plane of projection.

RULE.—Find first that projection of each point of shadow which is on the plane to which the cylinder is perpendicular, by noting the intersection of the like projection of a ray with the curve which is the like projection of the cylinder; then project this point upon the other projection of the same ray.

Any variations in the *form* of the abacus or column will change only the *form* of the shadow, not the manner of finding it.

EXAMPLES.—1°. Let the abacus be cylindrical, and then find the complete intersection of the whole cylinder of rays with the entire surface of the column.

Ex. 2°. Let it be hexagonal; octagonal.

Ex. 3°. Let the column be clustered.

[For convenience, the following, having the same solution, are added to complete the series.]

Ex. 4°. Let the abacus and column both be square.

Ex. 5°. Let the abacus be square or circular, and the column hexagonal.

Ex. 6°. Let the column in each of the preceding cases be perpendicular to **V**.

Ex. 7°. Let it be parallel to the ground line.

PROBLEM VII.—*To find the shadow of the upper base of a vertical hollow half cylinder upon the visible interior surface.*

*In Space.*—The solution is the same as in the last problem.

*In Projection.*—Pl. XVIII., Fig. 12. $ADH—A'H'$ is the half cylinder. The shadow is cast by a portion of the edge of shade $A—G'A'$, and by the portion, $AD—A'D'$, of the upper circle of the interior surface, which is also an edge of shade, since it separates the illuminated annular surface $AEHF$ from the darkened portion of the interior.

First, $A—G'G''$ casts the shadow $Aa$ on the base of the cylinder; then $A—G''A'$ casts the shadow $a—g'a'$, parallel to $A—G''A'$; and lastly, the shadow $a'c'D'$ is found by assuming points, as $CC'$ on $AD—A'D'$, and drawing rays, as $Cc—C'c'$, giving points, as $cc'$, where (181) $c$ is first found.

EXAMPLES.—1°. Let the cylinder be perpendicular to **V**.

Ex. 2°. Find the complete intersection of the entire cylinder of rays whose base is the whole upper circle, with the whole surface of the given cylinder produced upward. This will be an ellipse, of which $D'c'a'$ is an arc.

Ex. 3°. Let the cylinder be parallel to the ground line, with the plane $EF$ parallel to **V**.

Ex. 4°. In the last example, let the plane $EF$ be parallel to **H**.

PROBLEM VIII.—*To find the shadow of a vertical cylindrical turret upon a concave cylindrical roof surface.*

*In Space.*—The complete line of shade upon a material cylinder, limited by bases, consists, besides the elements of shade of its convex surface, of that portion of the circumference of each base which separates those parts of the base and convex surface, one of which is in the light and the other in the dark.

The intersections of the planes of rays tangent to the cylinder casting the shadow, with the one receiving the shadow, will be the shadows of the elements of shade (180) of the former. Their construction is an application of the general problem: "To find the intersection of a plane and cylinder." (D. G., Prob. XXXIII.)

The rays from all points of that portion of either base which is a line of shade form a cylinder of rays. The intersection of this cylinder with the cylinder receiving the shadow, is the shadow of that base, and is found as in (59).

*In Projection.*—Pl. XVIII., Fig. 8. Let $QEF$—$Q'E'F'$ be the turret, and let $ABCD$—$A'B'C'D'$ be the projections of a concave roof surface, whose elements are parallel to the ground line $CD$, and whose profile in the plane $OlO'$ is the arc $Ol$—$ll'$, shown after revolution about $lO'$ to the right, at $dl'$. Let $Oo$—$O'o'$ be the direction of the light. Then—

1°. *The shadow of an element of shade of the turret.*—The vertical line at $F$ is one of these elements, and $Ff$ is the horizontal trace of the plane of rays containing it, and is therefore (Prob. V., 1°) the horizontal projection of its shadow on the roof. To find the vertical projection of the same shadow, assume any element, as $bc$ of the roof surface, revolve $b$ to $b''$, whence it is projected to $b'''$, giving $b'''o''$ for the vertical projection of the same element. Then project $c$ upon $o''b'''$ at $c'$, the vertical projection of $c$. Other points being similarly found, and $I$, on the lower edge of the roof, being projected at $I'$, the vertical projection $I'c'h'$ of the shadow of an element of shade can be sketched. The shadow of the opposite element at $E$—omitted to avoid confusing the figure—can be similarly found.

2°. *The shadow of the upper base of the turret.*—The ray $Oo$—$O'o'$, through the centre of this base, pierces **H** at $o$, the centre of the equal geometrical shadow on **H**, the upper base. The problem then becomes: To find the intersection of the cylinder of rays $fnd$—$O$ with the roof cylinder. The light being taken at 45° (174), and crossing the second angle, both traces of the plane of rays *through* the axis

$Oo$—$O'o'$ and *parallel* to the axis of the roof-surface (59) coincide in $oo''$ parallel to $CD$, and $o''a$ ($o''al' = 45°$) is the revolved trace of this plane on the profile plane $OlO'$. Then assuming any trace, as $nn''$, parallel to $CD$, of a plane $A'n''p'''$, parallel to $oo''a$, we find $p'''q'$, and by counter-revolution (1°) $pq$, the two projections of the element of the roof contained in this plane, and $nN$—$n'F'$ that of the cylinder of rays contained in the same plane. These elements intersect at $qq'$ the shadow of the point $NF'$ upon the roof. Other points, as $hh'$, being similarly found and joined, give the shadow of the upper base of the turret upon the roof.

EXAMPLES.—1°. Bring the horizontal projection forward so as to bring $o$ in front of the ground line.

Ex. 2°. Change the direction of the light so that the traces as $oo''$ of the planes of rays shall not coincide.

Ex. 3°. Let the elements of the roof surface be oblique to **V**.

Ex. 4°. Substitute an inverted frustum for a cylindrical turret.

PROBLEM IX.—*To find the shadow of a square abacus upon a conical column, and the shade of the column.*

*In Space.*—Since the lines casting shadows are here straight, the problem becomes an application of the general one: "To find the intersection of a plane and a cone."

*In Projection.*—Pl. XVIII., Fig. 9. $FGM$—$G'E'$ is the abacus, $ABCDE$—$ACD'E'$ the column, and $B'R$—$V'R'$ a ray of light.

1°. *To find the shade of the column.*—To condense the figure, let $DHE$—$DEV'$ be a cone of the same axis and taper as the column. The horizontal trace of the ray $B'R$—$V'R'$ of the ray through its vertex is $R$, then $Rt$, tangent to $DHE$, is the horizontal trace of a plane of rays, tangent to this cone, and $tB'$—$t''V'$ ($t''V'$ not drawn) is one of the two elements or shade of the cone. $TQ$, parallel to $tSR$, is the horizontal trace of the parallel plane of rays tangent to the column, and $Tt$—$T't'$ is an element of shade of the column.

$QP'$, parallel to $SV'$, is the vertical trace of this tangent plane of rays.

2°. *The point of shadow in the meridian plane of rays FB'R.* This plane cuts from the column the element $ab-a'b'$, and from the edge $GF-G'$ of the abacus, the point $F$, $G'$; the ray $Ff-G'f'$ from which pierces $ab-a'b'$ at $ff'$, the required point of shadow, that of $F$, $G'$.

3°. *The shadow of the edge $GL-G'$.*—The vertical projection of this shadow falls on the vertical trace $G'g'$ of the plane of rays through the given edge. Any point of its horizontal projection is found essentially as $I'c'h'$ in Fig. 8 was; thus, assuming any point, as $m'$ or $f'$ on this shadow, project it, as at $m$ or $f$, upon the horizontal projection, as $AD$ or $ab$, of the element, as $AD'$ or $a'b'$, on which it is found.

4°. *The highest point, $kk'$, of the shadow of $GM-G'E'$.*—This point is the intersection of the plane of rays through this edge with the element $BH-B'H'$. Taking now $BB'V''$ as an auxiliary vertical plane of projection, $G''H''$, parallel to $BB'$, and at a height $HH''$, $=B'H'$, is the new vertical projection of the under surface of the abacus; $G''$ is that of $GG'$; $g''$, at a height equal to $g'y$, that of $gg'$; and hence $BH''$ is that of the element $BH$, and $G''g''$ that of the ray $Gg-G'g'$. Then $k''$, intersection of $G''g''$ and $BH''$, is the auxiliary projection of the point required, and $k$ and $k'$—making $B'k' = kk''$—are its required projections.

5°. *The point of shadow on the element of shade $Tt-T't'$.*—The plane $P'QT$ cuts the vertical plane through $GM-G'E'$ in the line $MI-M'I'$ parallel to $P'Q$, and thus (D. G., Prob. XI.) it cuts the edge $GM-G'E'$ at $I'I$. Then the ray through $II'$ meets $Tt-T't'$ at $ii'$, the required point of shadow. Any other points can be found by this method. For any secant from $R$ to $DHE-DE$ is the horizontal trace of a plane of rays containing elements of the cone $DHE-V'DE$. But as this cone is similar to and concentric with the column, and with its base equal to the upper base of the column, $B'xX$, for example, gives the element $xX$ of the column contained in a secant plane whose trace $Rx$, regarded as on the plane $D'E'$, cuts the edge $G'E'$ in the point whose shadow falls on $Xx$.

6°. *Indirect construction of points.*—This is here done by

the *method of one auxiliary shadow*, an application of (175, 3°), and stated thus. If any two surfaces, $M$ and $N$, intersect in any line $C$, the shadow on $N$ of a given line being the one most easily found, then where the shadow $l$, of $L$ on $N$, meets the intersection $C$, is a point of the shadow of $L$ upon $M$. In applying this often very useful method to the present problem, $M$ is the conical surface of the column; $N$ is any horizontal plane as $U''W''$; $C$ is the circle $U''W''-UrW$ cut from the column by this plane, and $L$ is the edge $GM-G'E'$ of the abacus. The ray $Gg-G'g'$, for example, pierces the plane $U''W''$ at $g'g$, giving $go$, parallel to $GM$, for the shadow ($l$) of $GM$ on this plane. Then $nn'$ and $oo'$, where this auxiliary shadow intersects the circle $UrW-U''W''$, are two points of the shadow of $GM-G'E'$ upon the column. By this method, therefore, points of shadow are found before knowing the points which cast them. The latter are, however, found by drawing rays inversely from $nn'$ and $oo'$, which will meet $GM-G'E'$ in the points casting these shadows. In the figure, the ray from $nn'$ will only meet $MG$ produced, showing $nn'$ to be an imaginary point.

EXAMPLES.—1°. Let the abacus be cylindrical.

Ex. 2°. Construct points of shadow on the column by means of secant planes of rays cutting it in elements.

Ex. 3°. Apply the method of (6°) to Prob. V. and to Prob. VIII.

**PROBLEM X.**—*To find the shadow of a vertical cylinder upon an upright cone.*

*In Space.*—This problem, as a practical one, might be: to find the shadow of a cylindrical chimney or turret upon a conical tower roof. As in Prob. VII., the line of shade of the cylinder consists of the elements of shade and of half the circumference of the upper base.

The shadows of the elements of shade are the intersections of the planes of rays containing them, with the cone. That of the edge of shade of the upper base is the intersection of the cylinder of rays containing it with the cone (58).

*In Projection.*—Pl. XVIII., Fig. 11. $CED-P'E'$ is the cylinder; $VAB-V'A'B'$ the cone, and $Oo-O'o'$ is a ray of light. The circle $oq$, equal to $CED$, and whose centre is the horizontal trace, $o$, of the ray $Oo-O'o'$ through the centre $OO'$ of the upper base of the given cylinder, is the base of the cylinder of rays through $CED-D'E'$. Proceeding as in (58), $R$ is the horizontal trace of the ray through the vertex of the cone. Hence any secant from $R$ to the bases of the *given cone* and the *cylinder of rays*, is the horizontal trace of a plane of rays cutting elements from both, whose intersections are points of shadow; provided that the elements of the cylinder of rays contain points of the semicircle $CED$ and that those of the cone are the ones first met by the rays.

Thus, the plane of rays whose horizontal trace is $RA$ cuts from the cone two elements, one of which is $AV-A'V'$, and from the cylinder of rays two elements, one of which, $aE-a'E'$, contains a point of $CED$. Then $ee'$, intersection of the ray $Ea-E'a'$ with the element $AV-A'V'$ of the cone, is a point of shadow on the cone. Other points, as $cc'$ and $dd'$, are similarly found.

The horizontal projections of the shadows of the element of shade at $D$, are $Ds$ on **H**, and $sd$ on the cone. The latter is a hyperbolic arc (36, 3°) whose vertical projection can be immediately found either by elements or by horizontal circles of the cone which shall intersect it between $ss'$ and $dd'$. (Prob. VIII., $cc'$. Prob. IX., $mm'$, $ff'$.)

EXAMPLES.—1°. Change the direction of the light.
Ex. 2°. Change the relative position of the bodies, keeping both upright.
Ex. 3°. Make the cylinder horizontal.
Ex. 4°. Make the axis of the cone horizontal. [In this example, and in 3, an auxiliary vertical plane perpendicular to the horizontal axis will be useful.]
Ex. 5°. Apply the method of Prob. IX., 6°.

### C—Shades and Shadows on Warped Surfaces.

182. We have under this head, general methods, based on general principles; and special ones, based on the properties of each surface. The former are here stated.

*Shades.*—Every plane containing an element of a warped surface is tangent to the surface at some point of that element (139). If then it be also a plane of rays, its point of contact will be a point of shade.

Otherwise: we have shown (159) how to pass a plane through a given line and tangent to a warped surface. If the line be a ray, the auxiliary parallel tangent cylinder there described will be a cylinder of rays, and hence its curve of contact, the curve of shade.

183. *Shadows on Warped Surfaces.*—The general method for finding these would be to pass planes of rays, each of which should contain an element of the surface, and should cut the line casting the shadow in some point (D. G., Prob. XI., or 54, 55). Then the ray through this point would meet the element in the shadow of the point.

This article and the last will be sufficient for the few cases in which these shades and shadows need be found.

Shades and shadows on the one warped surface of revolution (65) are found by methods which will be explained as applied to double-curved surfaces.

Other problems are sufficiently represented by the following problem of the screw.

184. *Description of the screw.*—Pl. XVIII., Fig. 13. Let the isosceles right-angled triangle $G'A'C'$, whose base $G'C'$ is vertical, uniformly revolve about the vertical axis $o$—$oo'$ in the same plane with it, and at the same time vertically and uniformly ascend a distance equal to $G'C'$, so as to occupy, after one such ascending revolution, the position $C'A''C''$. Every point of the triangle will thus generate a helix (100). The whole triangle will generate a solid called the thread of a triangular one-threaded screw. The upper and under surfaces of this thread are generated by $A'C'$ and $A'G'$ respectively, and are zones of oblique helicoids (127, 3°), the former comprised between the inner and outer helices generated respectively by $CC'$ and by $AA'$; the latter, between the helices generated by $CG'$ and $AA'$.

The circle of radius $oA$ is the horizontal projection of the outer helix, and the one of radius $oC$ is that of the inner helix.

The cylinder, of radius $oC$ and axis $o$—$oo'$, to which the thread is attached, is the core of the screw.

185. Any element is thus assumed. Produce $AC$—$A'C'$ till it meets the axis at $o'$. Then assuming $STo$ for example, project $S$ at $S'$ on the outer helix, and $o$ at $o''$, as far above $o'$ as $S'$ is vertically above $A'$, giving $S'T'o''$ for the vertical projection of $STo$; when, as a check upon the accuracy of the figure, $T'$, projected from $T$ on the inner helix, will also fall on $So''$.

PROBLEM XI.—*To find the shades and shadows on a triangular threaded screw whose axis is vertical.*

On account of the length of some of the topics in this comparatively complex problem, we will give the solution, "in space," and "in projection," for each separately.

1°. *A point of shade on the inner helix and underside of the thread by means of the declivity cone.*

*In Space.*—The tangent plane to a helicoid at any point of a given helix is determined by the tangent to the helix, and the element, at that point (138). The inclination of each of these lines to **H**—here perpendicular to the axis—is constant for all points of the same helix; hence the like is true for all tangent planes at points of the same helix. If, then, the line of declivity (Prob. VII.) of any one of these planes, cut from it by a meridian plane (33) of the screw, be revolved about the axis $o$—$oo'$ of the screw, it will generate a cone all the tangent planes to which will be parallel to corresponding tangent planes on the given helix. Hence a *tangent plane of rays* to the helicoid (182) will be parallel to the tangent plane of rays, easily found, to this declivity cone.

*In Projection.*—Pl. XVIII., Fig. 13. Making $Dd$ equal to the arc $Dj$, of the inner helix (102) $d$ and $b$, horizontal traces of the tangent, and of the element $BD—B'D'$, give $dbc$ the horizontal trace of the tangent plane to the under surface of the thread at $DD'$. Drawing $oc$ perpendicular to $db$, projecting it at $c'$, and producing $B'D'$ to $v'$ on $o'o$ produced, gives $oc—v'c'$ the generatrix of the declivity cone, whose horizontal trace is the circle $oc$, and to which the plane $db$ is tangent. All other like tangent planes to the under surface of a thread, at points on the inner helix, are parallel to tangent planes to this cone.

Now $cob$ is the constant angle between the line of declivity and element of contact of these tangent planes; hence, drawing $rh$ tangent to circle $oc$ from $r$, the horizontal trace of the ray $or—v'r'$, through the vertex $ov'$ of the declivity cone, it is the horizontal trace of the *plane of rays* tangent to this cone; hence, making $moh$ equal $cob$ and $om$ and $ob$ in like directions from $oh$ and $oc$, we have $om$ for the element of contact of a plane of rays parallel to the plane $rh$ and tangent to the inner helix at the point $mm'$ of the required curve of shade. The other tangent from $r$ to circle $oc$, not shown, will afford another point of shade.

Points of shade on the outer, or on any intermediate helix can be similarly found.

2°. *A point of shade on the outer helix and under surface by the method of helical transposition of a tangent plane.*

*In Space.*—Imagine a tangent plane to the helicoid at any point of the given helix. If it be also a plane of rays (182), the horizontal trace of a ray through its intersection with the axis will be in the horizontal trace of the plane. Otherwise this plane will intersect the cone generated by this ray in two elements, which may be considered as revolved positions of this ray. If then the tangent plane receive the same helical motion as the generatrix of the helicoid, it will continue tangent at successive points of the given helix, and when it has so revolved until either of the revolved rays, by an equal angular motion, again coincides with a ray, it will become a tangent plane of *rays*, when (182) its contact will be a point of shade.

*In Projection.*—Making $Aa = $ arc $AJ$, we have $au'$ the

horizontal trace of the tangent plane at $AA'$ to the under surface of the thread. This plane cuts the axis $o$—$oo'$ at $oe'$, where the element $A'u'$ in it meets the axis. The ray $of$—$e'f'$, through $oe'$, pierces **H** at $f$, not in the horizontal trace $au'$ of the tangent plane; which is thus *not* a plane of rays, but one that cuts the cone of vertex $oe'$ and horizontal trace $MfN$, generated by $of$—$e'f'$, in the elements $oM$ and $oN$. Now moving the plane $au'$ helically upward till $oN$ is again horizontally projected in $of$, it will then be a tangent plane of rays, and its point of contact, $AA'$, having an equal angular motion, $Aon = Nof$, will be found at $nn'$, another point of shade.

By this method two points on any helix can be found. For by helically revolving the plane till $oM$ coincides with $of$, we shall find a point of shade on the outer helix and on the back of the screw.

By reason of the uniformity of the screw, $m$ can also be projected at $m''$, etc., and $n$ at $n''$, etc.

EXAMPLES.—1°. Find *both* of the points derivable from the plane $db$ as $mm'$ was; and from the plane $au'$ as $nn'$ was.

Ex. 2°. Find points on an outer helix by (1°).

Ex. 3°. Find points on an inner helix by (2°).

Ex. 4°. Taking the light more nearly horizontal, find a point of shade on the *upper* surface of the thread, by each of the methods given in the problem.

3°. *The shadow of the curve of shade on the upper surface of a thread.*

*In Space.*—Points of this shadow are found by the method called *that of two auxiliary shadows*, which, by reference to Fig. $n$, can easily be stated as a general principle thus: If the shadows, $p$ and $q$, of two lines $P$ and $Q$ in space, upon a plane $M$, intersect as at $a$, the ray through $a$ must intersect both of the lines $P$ and $Q$ (it being the element common to the two intersecting cylinders of rays whose bases are $p$ and $q$, and whose elements are parallel), and $A$, where this ray intersects $Q$, will be the shadow on $Q$, of $A_1$, where the same ray intersects $P$.

*In Projection.*—Taking the horizontal plane $s't'$ ($= M$, Fig. $n$), in order to condense the figure, the shadow of

$mn-m''n''$ ($=P$, Fig. $n$) upon this plane is $pq$; that of any assumed element (185) $ST-S'T'$ ($=Q$, Fig. $n$), on which it is supposed that some point of the shadow of $mn-m''n''$ will fall, is $st$. Then drawing the ray $yUU''$, $y'U'U'''$ from $yy'$, where $pq$ and $st$ intersect, the point $UU'$, where it crosses $ST-S'T'$, is the shadow of the point $U''U'''$ where it meets the curve of shade.

4°. *The shadow of the outer helix on the thread below.*

*In Space.*—This shadow can be found as was the preceding, but for variety, it shall be found by direct construction by the general principle of (183).

*In Projection.*—Assume any element $FE-F'E'$ (185) on which a point of this shadow may fall. Its horizontal trace is $k$, and that of the ray $Ee''-E'e'''$ intersecting it is $e''$, hence $ke''$ is the like trace of a plane of rays through it.

The intersection of this plane with the outer helix could be found by (55), but by reason of the uniform helical motion of every point of the screw, a like helical revolution of the plane $ke''$ is better. Then revolve it thus till perpendicular to **V**, when its vertical trace will coincide with $E_1'o''''$, the like helically revolved position of $Eo-E'o'''$ (2°), found by making $EE_1 = l_1B$, where $l_1$ is on $ol$ perpendicular to $ke''$, and therefore the line of declivity of that plane. The plane $ke''$, thus made perpendicular to **V**, cuts the outer helix at $i'''$, thence projected at $i''$. Returning to the primitive position, $oB$ returns to $ol_1$; $oE_1$ to $oE$; $i''$ to $i$, by making $i'''i = E_1E = Bl_1$, and $i$ is then projected at $i'$. Thence the ray $ig-i'g'$ meets the element $EF-E'F'$, in whose plane of rays it is by construction, at $gg'$ a point of the required shadow.

Applying the method of (3°) to the helical arcs $S'B'$ and $i'''B'''$, we could find $x$ where the shadow leaves the screw. Or it could be found indirectly by joining $gg'$ with a point similarly found on an element, as $o''''E_1'$, produced below $E_1E_1'$.

The shadows on different threads are equal and similarly placed. Those not falling on the screw fall on the planes of projection, and are found as in Prob. I.

EXAMPLES.—1°. Find the shadow of the curve of shade—or the outer helix—upon the outer helix below by the method of (3°).

Ex. 2°. Find the shadow of a hexagonal nut upon the screw.

### D—Shades and Shadows on Double-Curved Surfaces.

186. The line of shade on a double-curved surface is wholly curved. Any point of it is imagined and constructed by *direct methods*, either as the *point of contact of a tangent plane of rays*, or of a *ray tangent to a section of the surface made by a secant plane of rays*.

Points of shade are often best found by indirect methods, which will be explained.

187. Points of *shadow* on a double-curved surface of revolution are in general easily found *directly* by constructing the intersection of rays through points casting shadows with the sections of the surface made by planes of rays through the latter points; also by the *indirect methods* of (Prob. IX., 6°) and (Prob. XI., 3°).

PROBLEM XII.—*To find the curve of shade upon a sphere.*

*In Space.*—This curve will be the great circle of contact of a cylinder of rays, which will also be one of revolution, since on a sphere the plane of the circle of shade, called the *plane of shade*, is perpendicular to the light.

By these principles, the shade as seen on each projection can be found separately.

*In Projection.*—Pl. XIX., Fig. 14. Let the circle of radius $OB$ be the vertical projection of a sphere, $V$ being the plane of the paper, and supposed to contain the centre of the sphere, and let $OL'$ be the vertical projection of a ray of light, taken at 45° (174). Then $L'C$ is the vertical trace of a plane of rays perpendicular to $V$ through the centre of the sphere (thence called the *perpendicular plane of rays*), and $AB$, perpendicular to $L'C$, is the like trace both of the plane of shade, and of that plane of rays which makes with **V** the same angle that the light does (thence called the *oblique plane of rays*). The two latter planes are perpendicular to each other and to the plane $L'C$.

If now the plane $L'C$ be revolved about its trace $L'C$ to coincide with **V**, the revolved traces of the other two planes upon it will appear as perpendiculars to each other at $O$.

Make $OL' = BF$, considered simply as the hypothenuse of the isosceles right-angled triangle $FOB$, and $L'L$ perpendicular to $OL$ and equal to $OB$, and $LO$ will be the revolved position of the ray through $O$, taken in the given direction (Prob. XII.), that is, of the trace of the oblique plane of rays. Then $OD$, perpendicular to $LO$, is the revolved trace of the plane of shade, and the projection of the curve of shade upon the plane $L'C$ (27, 8°).

Hence, drawing $ab$, $cd$, etc., planes parallel to $FC$, the circles of radius $Oa'$, $Oc'$, etc., cut by them from the sphere, give, by their intersections with $OD$, the auxiliary projection of the curve of shade, the points $ff'$, $ee'$, $DD'$ of the curve of shade, as may be more evident by considering the tangent rays parallel to $LO$ at $f$, $e$, $D$.

The vertical projection of the curve of shade, being an ellipse whose axes are $AB$ and $2OD'$, its three other quarters may be found from the quarter $Be'D'$.

The darkened portion of the figure represents the visible portion of the shade of the sphere.

EXAMPLES.—1°. Find the horizontal projection of the curve of shade in the same manner.

Ex. 2°. Change the direction of the light. [Both projections of a ray must then be given.]

## PROBLEM XIII.—*To find the curve of shade on a torus.*

Repeating the arrangement of topics employed in Problem XI., and for a like reason, we have—

DIRECT METHODS.—1°. *Points of shade on the apparent contours.*

*In Space.*—Pl. XIX., Fig. 15. The torus here meant is of the kind used in architectural mouldings, and is generated by a rectangle $O'O''C'D'$, to which a semicircle is tangent as at $C'A'D'$, this compound figure revolving about the side $O'O''$. The contours are then those forming the projections, viz., the greatest *parallel* (Prob. XLIX.) $ATB$—$A'B'$, and the *meridian* in or parallel to **V**.

The points of shade on these contours are the points of contact of tangent planes of rays, perpendicular respectively to **H** and **V**.

*In Projection.*—$RO$—$R'O'$ being the projections of a ray, the tangent at $T$, parallel to $RO$, and at $a'$ and $b'$, parallel to $R'O'$, are the traces of the planes just mentioned, and give the required points $TT'$, and $a'a$ and $b'b$ (186).

2°. *The highest and lowest points.*

*In Space.*—These points, the axis of the torus being vertical, are the points of contact of rays of light with that meridian which is in a meridian plane of rays.

*In Projection.*—$RO$ is the horizontal trace of the meridian plane of rays. Revolving it about $O$—$O'O''$ into **V**, its meridian coincides with that of the vertical projection. The ray $RO$—$R'O'$ appears after the same revolution at $R''O$—$R'''O'$. Then radii, as $o'r'''$, at $o'$ and $n'$, centres of the semicircular ends of the meridian, and perpendicular to $R'''O'$, give the revolved points of contact as $r'''r''$ of rays. By counter-revolution, $r'''r''$ returns to $rr'$, the lowest point. The highest point is similarly found.

3°. *Intermediate points by tangent planes of rays.*

*In Space.*—Any plane containing a ray is a plane of rays. If, then, a plane be passed through a ray, and perpendicular to any meridian plane of the torus, a tangent to the meridian in that plane, parallel to the trace on the latter of the perpendicular plane, will be the trace of a tangent plane of rays perpendicular to the meridian plane. Hence by (80) the contact of this tangent and meridian will be the point of contact of a tangent plane of rays with the torus, and hence a point of shade.

*In Projection.*—Assume any meridian plane $Oc$. $OQ'$ is its own projection upon this plane. $RR'$ is there projected by the line $RR_1$—$R'R_1'$. Revolving the meridian plane about $O$—$O'O''$ into **V**, the projected ray appears at $OR_1''$—$O'R_1'''$, when $o'c'''$ perpendicular to $O'R_1'''$ gives $c'''c''$ the revolved, and thence, as in (1°) $cc'$, the primitive position of the point of contact of a tangent plane of rays perpendicular to the meridian plane $Oc$.

Other points can be similarly found.

INDIRECT METHODS.—1°. *By auxiliary circumscribed cones.*

*In Space.*—This method stated as a principle applicable of all double curved, and the warped (Theor. V.) surface of revolution, is as follows:

If two surfaces, $M$ principal, and $N$ auxiliary, are tangent to each other on a circle of contact (as a cone and sphere may be), then the intersection of the curve or plane of shade of $N$ (supposed to be easily found) with the circle of contact of $M$ and $N$, is a point of shade common to both, and thus a point of the required curve of shade of $M$.

*In Projection.*—Pl. XIX., Fig. 16. Representing the torus of Fig. 15 again, to avoid confusion of lines, assume $M'N'$—$MN$ anywhere between the highest and lowest points of shade, as the circle of contact of a cone whose axis is $O$—$O'V'$, that of the torus, and whose generatrix is $M'V'$, tangent at $M'$ to the meridian $A'M'B'$ of the torus. Then $n$ is the trace of the ray $On$—$V''n'$ upon the plane $M'N'$, and consequently $nt$ is the trace, on this plane, of the tangent plane of rays to the cone, and $Ot$—$V't'$ (not necessarily drawn) is the element of shade of this cone. Then $tt'$, intersection of this element of shade of the cone, with the circle of contact $MN$—$M'N'$ of the cone and torus, is a point of shade on the torus.

2°. *Auxiliary tangent spheres.*

*In Space.*—This solution is the same in principle as the last, only instead of actually finding the circle of shade of the sphere, we only find where its plane cuts the circle of contact of the sphere and cone.

*In Projection.*—Assume, as before, $K'L'$—$KqL$ as the circle of contact of a sphere with the torus. The centre

$OO'$ of this sphere will then be where $K'o'$, through the centre of the semicircular portion $M'A'D'$ of the meridian meets the axis $O-O'V'$ of the torus; and the circle $O'K'$ is the vertical projection of the sphere.

Now $O'Q'$, perpendicular to $V'n'$, is the vertical trace of the plane of shade of the sphere (D.G., Theor. II.). Projecting $Q'$ at $Q$, the line $Qq$, perpendicular to $On$, is the trace of the same plane on the plane $K'L'$. Hence $qq'$, intersection of $Qq-Q'q'$ with $KqL-K'L'$, is the intersection of the plane of shade, and hence of the curve of shade of the sphere with the circle of contact of the sphere and torus, and hence is a point of shade on the torus.

If the axis of the torus $O-O'O''$ were not in **V**, the meridian plane parallel to **V** would have been used just as **V** itself has been here.

EXAMPLES.—1°. Let $O-O'O''$ be in space, and find the complete curve of shade.

Ex. 2°. Given both projections of a sphere, find its curve of shade as in the torus. [Here the given and auxiliary spheres will coincide.]

Ex. 3°. Find the curve of shade on the inner or concavo-convex portion of an annular torus (73), when its axis is vertical. See Fig. 18.

Ex. 4°. Find the curve of shade on the inner or concavo-convex portion of an annular torus (73), when the axis is perpendicular to **V**.

Ex. 5°. Find the curve of shade on an ellipsoid of revolution, whose axis of revolution is vertical, by the various methods of the last problem.

Ex. 6°. Find the curve of shade on an ellipsoid of revolution, whose axis of revolution is vertical, by means of an auxiliary vertical plane of projection, $V_1$, parallel to the light [on which, since the curve of shade is plane, it will be projected in a straight line joining the points of contact of tangents parallel to the projections of rays on $V_1$].

PROBLEM XIV.—*To find the shadow of a niche on itself.*

*In Space.*—A niche, as usually formed, is a hollow in a wall, composed of a vertical semi-cylinder of revolution capped by a quarter sphere, generated by the quarter revolution of the upper base of the cylindrical portion about its diameter.

The total shadow is in four parts: that of the vertical

edge of the cylindrical part upon its lower base; that of the same upon the cylindrical surface; that of the vertical semicircle of the face upon the cylindrical part; and, finally, that of this semicircle upon the spherical part.

The three former are easily found as in previous similar problems. The latter is the intersection of the cylinder of rays through the first semicircle, with the spherical surface and would be an equal semicircle, in case of a full hemisphere. For if parallel rays, $AD$, $CB$, etc., Pl. XIX., Fig ($n$), are passed through all points of any great circle, as $AB$, of a sphere, but oblique to its plane, all but the two which, like $mn$, are tangents, will be secants containing chords, as $AD$, $ad$, etc., of the sphere. But the plane $PQ$, perpendicular to these chords and through the centre of the sphere, will bisect them all. Hence $AB$ and $CD$, being thus symmetrical with $PQ$, are alike great circles; which intersect in a common diameter at $O$.

The method of (Prob. IX., 6°) is also readily employed.

*In Projection.*—Pl. XIX., Fig. 17. $AbB—A'B'C'D'$ is the cylindrical, and $AOBb—A'E'B'O'$ the spherical part of the niche, the face of which is parallel to **V**; and $RL—R'O'$ is the direction of the light.

1°. *By direct construction.*—Taking the ray $RL—R'O'$, its trace on the upper base of the cylinder is $L, O'$. Hence making $O'L'' = OL$, and perpendicular to $R'O'$, we have $R'L''$ the revolved position of the ray $RL—R'O'$ about the trace $RO—R'O'$ of a plane of rays through it and perpendicular to the face of the niche. The like revolved position of the semicircle of the sphere in the plane $R'O'$ is $R'T'B'$, hence $a''$, its intersection with the revolved ray $RL''$, is the revolved position of the point of shadow, which, by counter-revolution, is found at $a'$. Again assume $m'p'$, parallel to $R'O'$, as the trace, upon the face of the niche, of a plane of rays, perpendicular to that face, and hence to **V**. This plane cuts from the spherical part the semicircle whose vertical projection is $m'p'$, and whose position, after revolution about the trace $m'p'$ into the face, is $m'l''p'$. Then $m'l''$ parallel to $R'L''$ is the revolved ray at $m'$, and $l''$

is therefore the revolved shadow of $m'$, which, by counter-revolution, returns to $l'$. The horizontal projection $l$, of $l'$, is found by making $Tl$, on $ll'$, equal to $l'l''$.

Other points may be similarly found. At $TT'$, contact of a tangent plane of rays perpendicular to **V**, the points casting and receiving shadow unite, then $T'l'a'$ is the vertical projection of a half of the semi-ellipse of shadow, all below $A'B'$ being imaginary.

2°. *The indirect method.*—Assume any plane $FK$ parallel to the face of the niche. It cuts from the spherical part the semicircle $FK-F'n'K'$, and from the cylinder of rays through the front semicircle, the semicircle equal to $A'E'B'$, and whose centre is $MM'$, where the plane $FK$ meets the ray $OM-O'M'$, axis of this cylinder. Then drawing an arc $n'y'$ with centre $M'$ and radius $O'A'$, its intersection $n'n$ with $FK-F'n'K'$, is (Prob. IX., 6°) a point of the shadow on the spherical surface.

3°. *The point of shadow on* $AbB-A'B'$.—This is found with practical accuracy by indirect means; also by an easy, direct construction.

Thus, suppose the spherical part produced below the plane $A'B'$, as in finding $a'$, the intersection $s's$ of the quarter ellipse $T'a'$ with $A'B'$ gives the point desired. Also, suppose the cylindrical part to be produced upward, and find the shadow, from $hh'$ upward, of the semicircle $AB-A'E'B'$ (181) upon this surface. The intersection of this shadow with $A'B'-AbB$ will again give the same point $s's$.

But again: $O'T'$, perpendicular to $R'O'$ (Theor. II.), is the trace, on the face of the niche, of the plane of the circle of shadow; hence $l'k'-lk$, where $l'k'$ is parallel to $O'T'$, is (27, 5°) a line in this plane. Therefore, $k'k$ is a point of the trace of this plane on the upper base of the cylinder, and $OO'$ is another. Then $Ok$ is that trace, and $ss'$, its intersection with $AbB-A'B'$, is the point of shadow on that circle.

EXAMPLES.—1°. Substitute for the niche a hollow hemisphere with visible interior and its great circle parallel to **V**.

Ex. 2°. In (1) let the great circle be parallel to **H**.

SHADES AND SHADOWS. 221

Ex. 3°. Let $AbB$ be a semi-ellipse, and substitute for the quarter sphere a quarter ellipsoid of revolution having $AB—A'B'$ for its longer axis. [Use the indirect method, and an auxiliary plane of projection perpendicular to the ground line.]

PROBLEM XV.—*To find the brilliant point of a double-curved surface.*

*In Space.*—The *absolute* brilliant point of a surface is the one which receives the most light—that is, the one at which a ray is normal to the surface.

The *apparent* brilliant point, relative to any given observer, is the one at which the most light is reflected to his eye in accordance with the principle that the incident and reflected rays make equal angles with the reflecting surface. Vertical projections being the ones generally shaded in a finished drawing, since they seem more like the reality, the reflected rays mean those which are perpendicular to the vertical plane of projection.

The brilliant point is then the one at which a ray of light and a perpendicular to **V** make equal angles with the reflecting surface—that is, with a *tangent plane* to it, and hence also with a *normal* to it.

*In Projection.*—1°. *On the sphere.*—Pl. XIX., Fig. 14. The vertical projections of the incident and reflected rays are respectively $L'O$ and $O$. Their revolved positions about the trace $FC$, into **V**, are $LO$ and $OB$; hence $p$, where the bisecting line $Op$ of the angle $LOB$ meets $FBC$ the revolved great circle $FOC$, is the revolved position of the brilliant point. By counter-revolution, $p'$ is its vertical projection.

2°. *On a concave tower roof.*—Pl. XIX., Fig. 18. $A'CB'—A'O'B'$ is the half roof, **V** being taken through its axis; $OR—OR'$ is a ray of light, and $OC—O$ the direction of reflected rays.

Revolve the plane $R'OC$ about $OC$ into **H**, when $OR—OR'$ will appear at $OR''—OR'''$, and $N''O$, bisecting the angle $R''OC$, is the revolved position of the bisector of the angle between the incident and reflected rays at $O$.

By counter-revolution, $ON''$—$ON'''$ appears in $R'OC'$ at $ON'$—$ON$. Once more, revolving $ON$—$ON'$ about the vertical axis $OO'$ and into **V**, it appears at $ON_1$—$ON_1'$. Now $Q'$ being the centre of the roof arc $A'O'$, draw $Q'q'''$ parallel to $ON_1'$, and project $q'''$ at $q''$, and $q''q'''$ will be the revolved position of the brilliant point. By equal counter-revolutions of $ON_1$—$ON_1'$ and $q''q'''$, the latter returns to $qq'$, its real position.

Hence, in making a finished shading of the vertical projection, $q'$ would be the lightest point of the shading, from which the latter, very light at first, would be made gradually darker until it reached the line of shade.

188. *To find the brilliant point on an upright screw*, find the point of contact of a tangent plane which is perpendicular to the line bisecting the angle between a ray of light and a perpendicular to **V**.

The brilliant *point* is replaced on surfaces with which a tangent plane has an element of contact, by a brilliant *line* or element. But in most cases, as in that of an upright cylinder or cone, there will be no tangent plane perpendicular to the bisecting line just described (see $NO$—$N'O$, Fig. 18), and it will be practically sufficient, in the case supposed, to take the brilliant element in that vertical plane whose horizontal trace bisects the angle between the horizontal projections of an incident and reflected ray through a point on the axis of the body.*

EXAMPLES.—1°. Find the brilliant point on an ellipsoid.
Ex. 2°. On a torus.

* For numerous other examples, including those with divergent rays of light, see my volume on SHADES AND SHADOWS.

# LINEAR PERSPECTIVE.

### FIRST PRINCIPLES.

189. Hitherto the eye has been supposed to be at an infinite distance (5) from each of the planes of projection, and the projecting lines (5) to be perpendicular to them.

We shall now consider the case in which the eye is at a finite distance from the plane of projection, so that the projecting lines will radiate from the eye to the determining points of the object.

This case corresponds to the natural conditions under which objects are actually seen. The projecting lines are then the rays remitted from points of the object to the eye, and are thence called *visual rays*, and the system of projection is that described in (6, II.).

190. PERSPECTIVE, then, *descriptively defined*, is the representation of any object, upon some surface, usually plane, by a figure that shall appear to the eye situated at a certain point, just as the object itself would when seen from the same point.

191. Such representation embraces two branches: first, the construction of the correct *form* of the perspective

figure; second, the addition of proper shading and coloring.

The former branch, only, concerns us now, and is called *linear perspective*. The latter, or the representation of natural effects of shade and color by the rules of art, is *aërial perspective*.

192. Any plane passing through the point of sight is called a *visual plane*, all lines in it and containing that point being *visual rays*.

A plane can always be passed through a given point and straight line, hence a visual plane can always be passed through any given straight line, and as it will contain the visual rays from all points of that line, its trace on the perspective plane will be the perspective of that line. If the line be parallel to the perspective plane, *its perspective will evidently be parallel to the line itself.*

193. The visible boundary of an object is called its *apparent contour*, and in the case of any curved surface, it is the curve of contact of a circumscribed tangent cone, whose vertex is the eye.

194. LINEAR PERSPECTIVE, therefore, *geometrically defined*, consists in finding *the intersection of the perspective plane with the visual cone* (or pyramid) whose vertex is at the eye or *point of sight*, whose elements (edges) are visual rays, and whose base is the apparent contour of the given object.

Problems in perspective being thus, including the preliminary construction of apparent contours, either problems of tangency, or of the intersection of a plane and cone, the subject is obviously an application of descriptive geometry.

195. The angle, at the eye, subtended by any object, is called the *visual angle*, and its size determines the *apparent* size of the object. Unless otherwise indicated, this angle will be understood as the *greatest horizontal angle* subtended by the object.

## A—Perspectives of Plane-sided Bodies—Various Methods.

PROBLEM I.—*To find the perspective of a monument composed of a square prism, capped by a square pyramid.*

*In Space.*—If threads were stretched from the principal points of the object to a point representing the position of the eye, the figure formed by joining the points in which these threads would pierce the perspective plane would (189) be the perspective of the object. Models illustrating the definition (194) of perspective are actually made in this way.

But perspective figures could not generally be thus constructed. We substitute for the actual object its two projections which (13, second) are equivalent to it, likewise for the eye and the visual cone their equivalent projections, and for the perspective plane its traces.

*In Projection.*—Pl. XX., Fig. 1. Let $ABDV—A'G'C'V'$ represent the given monument, $EE'$ the position of the eye, and $AA'Y$, perpendicular to the ground line $B'A'O$, the perspective plane having the same position.

The perspective figure in $AA'Y$ will appear by revolving that plane about its vertical trace $A'Y$ till it coincides with **V**; but as such a revolution would confound together the perspective and the vertical projection, the perspective plane is first translated to the new and parallel position $XOZ$, at any sufficient distance from $AA'Y$, before revolution.

Again: remembering that the spectator at $E$ faces the perspective plane $AA'Y$ perpendicularly—that is, in the direction $Ep$—the point $B$, for example, is at his right. It must so appear in the perspective figure, as at $b$, in order not to reverse the figure, hence the new position $XOZ$ of $AA'Y$ must be revolved to the left about $OZ$ to coincide with **V**.

These preliminaries, alike for all cases under the present method, being explained, then, for example, $VE—V''E'$ is the visual ray from the vertex $VV'$. It pierces the perspective plane at $pp'$, perspective of $VV'$, which is thence translated along $pp_1$—$p'p_1'$, parallel to the ground line, to $p_2p_1'$ on $XOZ$. Thence, by the revolution of $XOZ$, already described, $p_2p_1'$ describes the horizontal quadrant $p_2p_2$—$p_1'v$, giving $v$ the desired perspective of $VV'$.

Likewise $b$ is the perspective of $BB'$, as may be traced by the full lettering of its construction, and all the other points may be similarly found.

Having taken $VE$ perpendicular to $AA'Y$, to represent the observer as directly in front of the object, $EE'$ is projected on $XOZ$ at $p_1e'$, which appears after revolution at $E''$. If then the perspective figure be viewed by placing the eye on a perpendicular to the paper at $E''$, and at a distance from it equal to $Ep$, the figure will perfectly represent the given object, and will exactly do so at no other position of the eye.

EXAMPLES.—1°. Let the pyramid be octagonal.
Ex. 2°. Replace it by a hexagonal prism.

*Remarks on the Method of Three Planes.*

196. 1°. *Its extreme simplicity.*—It obviously conforms, in the construction of every point, to the most elementary definition (194) of perspective, so that one point having been found, as $v$, nothing remains but to repeat the same process for every other point required.

This method is therefore peculiarly appropriate to those who are making constant or frequent use of perpendicular projections, but who comparatively seldom need to make perspective constructions.

2°. *The apparently large amount of paper covered.*—But in practice, let $UXYZ$ represent a sheet of drawing-paper to be occupied by the perspective figure only. Suppose this sheet to be stretched on a large drawing-board or table; the auxiliary projections would then be pencilled, only, on cheap paper and temporarily fastened at the side of the drawing-paper, as indicated by $RQYD$. Besides, on becoming accustomed to the method, the operations can be greatly condensed by supposing the given object to be in the second angle (7) which will transfer all the construction lines now in front of the ground line $B'O$ to the opposite side of $B'O$, where they might in practice be distinguished by colored lines. Moreover, the perspective plane $AA'Y$ can then be revolved to the *right* about $A'Y$, into **V**, without translation.

3°. *The convergence of lines*, as $ab$ and $gc$, which in space are parallel, but are seen obliquely. The *correct relative direction of lines* in a perspective figure being of prime importance to an undistorted figure, it will be useful to construct the points to which such lines converge. This is easily done as in the next problem.

PROBLEM II.—*To construct the perspective of a pier, with the points of convergence of the perspectives of its parallel lines.*

*In Space.*—The dispositions in space, as indicated by the figure, being the same as in the last problem, if a visual ray be drawn parallel to any line of the given object, it will meet that line only at infinity. The intersection of this ray with the perspective plane will, as in any other case, be the perspective of the point from which the ray proceeds—that is, of a point at infinity. But a visual ray, $R$, which is parallel to one of several given parallels, $L$, $L_1$, $L_2$, etc., will be parallel to them all, and will meet them all at one and the same infinitely distant point, $I$; hence the intersection of this ray with the perspective plane will be the perspective $i$ of that common infinitely distant point $I$. Hence the perspectives $l$, $l_1$, $l_2$, of the parallels $L$, $L_1$, $L_2$, etc., will meet at $i$ perspective of the point $I$ common to all the parallels.

*In Projection.*—Pl. XX., Fig. 2. $ACFM—A'C'F'M'$ represents a stone pier, whose sides, $AC$ and $FM$, are parallel, and whose top is partly level, and at $BCFK—B'C'G'K'$ partly inclined to **H**.

The perspective figure, as illustrated in full by $a$, perspective of $AA'$, may be wholly found as in the last problem. But now let $ER$—$E'R'$ be that visual ray which is parallel to $CB$—$C'B'$, and to all lines parallel to the latter line. This ray therefore meets $CB$—$C'B'$ and all parallels to it at infinity. It also meets the perspective plane $CD'R_1'$ at $RR'$, which is thence translated to the new position $XOR_1'$ of that plane, at $R_iR_1'$, and after revolution as before about $OR_1'$ into **V**, appears at $V$. Hence $V$ is the perspective of the point at infinity where $CB$—$C'B'$ and all parallels to it meet each other, and $cb$ and $gk$, perspectives of $CB$—$C'B'$ and $FK$—$G'K'$, therefore meet at $V$.

In like manner, $ER$—$E'r'$ is the visual ray which is parallel to all horizontal lines which are parallel to $BA$—$B'A'$. This ray pierces the perspective plane at $Rr'$, which is translated to $R_ir_1'$, and revolved in the arc $R_iR_2$—$r_1'H$ to $H$, the perspective of the infinitely distant point at which $BA$—$B'A'$ and all parallels to it meet. At $H$, therefore, the indefinite perspectives, as $dh$ and $ba$ of such parallels, will meet.

Having $V$ and $H$, also $c$ and $a$, the point $b$ is the intersection of $cV$ and $Ha$, thus found without the visual ray from $BB'$. Also $a$, for example, might have been found as the intersection of $bH$ (supposing $b$ already known) with either $n_2a$, perpendicular, or $n'a$, parallel to the ground line.

197. The points $V$ and $H$, and others which might be similarly found for other sets of parallels, are no others than those commonly called *vanishing points*. They are obviously useful, and are found by the following,

RULE.—*To find the vanishing point of any system of parallel lines*, find the intersection of the visual ray parallel to them with the perspective plane. This intersection will be the vanishing point in which the perspectives of all those parallels will meet.

*Note*, that in like manner the trace of any visual *plane* upon the perspective plane is the perspective of its infinitely distant intersection with any parallel planes, and is thence called the *vanishing line* of those planes.

198. The perspectives of shadows can be found from their projections, just as the perspectives of the objects casting them are. It is, however, necessary to remember that, in making the perspective, the observer is facing the perspective plane, and that the light therefore should have the same position relative to the horizontal and the perspective planes that it usually has, relative to the two planes of projection (178).

PROBLEM III.—*To find the perspective of a square block and of its shadow on the horizontal plane.*

*In Space.—By visual rays.* Having the shadow itself, the visual ray from any point of it will pierce the perspective plane in the perspective of that point.

*By vanishing points*, shadows parallel to the lines casting them will have the same vanishing points as those lines, and may be partly found on this principle.

*In Projection.*—Pl. XX., Fig. 3. *The block.* $ABCD—A'B'G'$ represents the block. Combining the methods by visual rays and vanishing points, make $Of = F'A$; and $fa$, equal and parallel to $F'A'$, is the perspective of $F'A'$, this line being in the perspective plane. Also make $On_i = F'n$, and $Or_i = F'r$, and the perspectives of the vertical edges at $B$ and $D$ will be in perpendiculars to the ground line at $n_i$ and $r_i$.

But, now, $ER—E'R'$, the visual ray parallel to $AB—A'B'$, pierces the perspective plane at $RR'$, which, by translation and revolution, as in Fig. 2, making $XV = FR$, gives $V$, the vanishing point (197) of $AB—A'B'$ and of all parallels to it. Similarly, $S_i$, vanishing point of $AD$, etc., is found, $XS_i$ being made equal to $FS$, where $S$ indicates the intersection of $ES$, parallel to $AD$, with $R'F'r$.

This done, $fV$ and $aV$, $fS_i$ and $aS_i$ are the indefinite perspectives of $AB—F'G'$ and $AB—A'B'$, $AD—F'H'$ and $AD—A'D'$, respectively. They are then limited at $g$ and $b$, and $h$ and $d$, by the perpendiculars $n_i b$ and $r_i d$.

*The Shadow.*—$AA_i—A'A_i'$ is the proper direction of the

light according to (174) and (178), giving $A_1A_1''$, for example, as the shadow of $AA'$. Then, as partly shown by the ray $A_1E$, translating $q$ to $q_1$ and revolving it to $q_2$, the shadow $a_1$ of $a$—perspective of the real shadow $A_1A_1'$—is found as all the points of Fig. 1 were. Thence, as $A_1B_1$ is parallel to $AB$, its perspective $a_1b_1$ will have the same vanishing point $V$, as $ab$, and will be limited at $b_1$ by $p_2b_1$. Likewise the shadow of $bc$ proceeds from $b_1$ towards $S_1$.

199. The line from a point to its shadow is a ray of light; hence $aa_1$ and $bb_1$ are the perspectives of the rays $AA_1$—$A'A_1'$ and $BB_1$—$B'B_1'$. These lines could have been found first, and then used in finding $a_1$ and $b_1$, by finding their vanishing points by the rule (197). Likewise $fa_1$ and $gb_1$, perspectives of $AA_1$—$FA_1'$ and $BB_1$—$G_1B_1'$, the horizontal projections of the same rays, could have been found.

EXAMPLE.—Find $a_1$ and $b_1$ as just indicated.

PROBLEM IV.—*To find the shadow of a shelf on a lateral wall, perpendicular to the perspective plane.*

*In Space.*—The vertical plane of projection is such a plane, relative to the perspective plane. The perspective of the shadow on **V** can then be found either wholly by the method of (Prob. I.), or by the principle of (199).

*In Projection.*—Pl. XX., Fig. 4. $AHBD$ is the horizontal, and $A'C'$ the vertical projection of the shelf, and $AFQ$—$D'F'Q'B'$ is its shadow on **V**. The perspective of the shelf is found as shown, as in Fig. 1; the end $aod$, however, needing only immediate translation and revolution, since it coincides in space with $AD$—$A'D'$ the end of the shelf which is in the perspective plane. The point $q$ of the shadow is also found by the visual ray $QE$—$Q'E'$, as shown.

*Otherwise, by vanishing points.*—The visual ray $ER$—$E'R'$, parallel to the rays of light, meets the perspective plane at $RR'$, vanishing point of rays of light (197), which point, by translation and revolution, making $OR_1 = AR$, appears at $L$.

Likewise $Er$—$E'R'$ the visual ray parallel to $D'F'$—$AF$, the two projections of the vertical projection of a ray of light, pierces the perspective plane at $rR'$, which, by like translation to $r_1R_1'$, and revolution, appears at $S$, the vanishing point of the projections of rays upon a vertical plane **V**, perpendicular to the perspective plane. Then as the point of shadow $FF'$, for example, is the intersection of the ray $DF$—$D'F'$ with its own vertical projection $AF$—$D'F'$, the perspective $f$, of $F$, is the intersection of $dL$, perspective of the ray from $DD'$ with $oS$, perspective of the projection on **V** of the same ray. Thus $ofqth$ is the shadow required.

The visual ray $QE$—$QE'$ gives $k$, the perspective of the point $Q$ of the ground line, and thus $Ok$ as the perspective of the base line $AQ$ of the lateral wall **V**.

EXAMPLE.—Substitute for **V** any vertical plane oblique to the perspective plane. [$Er$ will be parallel to the base line of the new plane.]

### Indirect or Artificial Methods.

· 200. The method of three planes and visual rays, only, may be called the direct and natural method of perspective; *direct*, because no auxiliary perspectives are employed in finding required perspectives; *natural*, because it obviously conforms immediately to (190) the simplest definition of perspective.

Yet simple as this method is, especially, as we shall see, in its application to double-curved surfaces, other methods are in very general use, varying both as to the *system of construction lines adopted*, and the *selection and management of the principal planes employed*. We shall next explain some of these methods, partly as an aid in consulting other works, and partly because each method is best under certain conditions.

201. *Systems of construction lines.*—These all rest on this principle, viz.: If any two lines $L$ and $L_1$ pass through a point $A$, the intersection, $a$, of their perspectives $l$ and $l_1$ will be the perspective of $A$. For the visual planes (192)

containing $L$ and $L_1$ will intersect each other in the visual ray $AE$ from $A$ to the point of sight, $E$. Hence the intersection of this ray with the perspective plane, and which is the perspective $a$ of $A$, will be the intersection of the traces of these visual planes with that plane—that is, at the intersection of the perspectives $l$ and $l_1$, of the lines $L$ and $L_1$.

This method is *indirect* and *artificial* (200), because $a$ is found by the substitution of two auxiliary perspectives, $l$ and $l_1$, for the visual ray $AE$.

202. The perspectives of the auxiliary lines chosen are found on the following principle:

*The indefinite, or unlimited perspective of a line is determined by its trace on the perspective plane, and its vanishing point.* The latter point is (197) simply the perspective of the infinitely remote point of the line; the former is obviously a point in the perspective of the line, since the visual ray from that point meets the perspective plane at that point.

203. It is convenient that the trace of an auxiliary line should be at the same height as the point whence it proceeds, hence in practice pairs of *horizontal* auxiliary lines are preferred. Finally, therefore, the systems of auxiliary lines actually chosen are chiefly two.

*First. Lines of the given object,* when, as in regular buildings, these mostly occur in parallel sets.

*Second.* The lines commonly called *diagonals and perpendiculars,* and which are thus defined:

*A diagonal* is a *horizontal* line which makes an angle of 45° with the perspective plane.

*A perpendicular* is perpendicular to the perspective plane.

204. *Auxiliary planes.*—Besides auxiliary *lines,* we may pass *any visual plane,* together with *one* auxiliary *line,* through a given point. The perspective of this point will then be the *intersection of the perspective of the auxiliary line,* with the trace of the auxiliary visual plane upon the perspective plane.

This visual plane is usually taken perpendicular to a plane of projection. When vertical, its vertical or perspective trace is perpendicular to the ground line. When perpendicular to the perspective plane, its like trace is identical with the vertical projection of a visual ray.

205. The trace, on the perspective plane, of a *horizontal* visual plane is called the *horizon*, and, by (197, *note*), it is evidently the perspective of the indefinitely distant natural horizon, and is the *vanishing line of all horizontal planes;* and hence contains the *vanishing points* of all horizontal lines.

206. In the use of two planes, **V** being also the perspective plane, the given objects are almost invariably placed in the second angle, not of necessity, but so as to reduce in the perspective any instrumental errors in the projections, the perspective figure being then necessarily smaller than the projections, and therefore, moreover, more sure to be contained within the assigned limits of the drawing.

207. In these indirect methods, the horizontal projection $E$ of the eye is often called the *station point;* its vertical projection, $E'$, the *centre of the picture;* and the vanishing points of diagonals, *points of distance;* since by (197, 203) they, and the eye itself, are obviously at equal distances from the centre of the picture.

From (197, 205) we now have, relative to important particular lines:

1°. *The vanishing point of perpendiculars is at the centre of the picture.*

2°. That of *diagonals* is in the *horizon*, at the same distance from the centre of the picture that the eye is.

3°. That of *any horizontal line* is in the horizon.

4°. That of any *parallel to the perspective* plane is at infinity; whence we have again the final principle of (192).

5°. That of *any line at* 45° with the perspective plane is on a circumference struck from the centre of the picture, with a radius from that point to the points of distance.

## 234 LINEAR PERSPECTIVE.

**PROBLEM V.**—*To find the perspective of a horizontal and of an oblique straight line; first, by visual rays; second, by trace and vanishing point; the vertical plane being also the perspective plane.*

*In Space*—The horizontal line is oblique to **V**, and the other one is oblique both to **H** and **V**. Both **H** and **V** retain their primitive positions throughout the construction.

*In Projection.*—Pl. XXI., Fig. 5. 1°. *The horizontal line.* Let $AB—C'D'$ be this line, in the second angle, and let $EE'$ be the place of the eye.

*By visual rays.*—$AE$ is the horizontal projection of the visual ray from the extremity $A,C'$ of the line, and $C'E'$ is its vertical projection. Its trace (202) on the perspective plane, **V**, is $c$, which is therefore the perspective of the point $A,C'$. Likewise, and briefly, naming lines by their projections (13) $BE—D'E'$ is the visual ray from the point $B,D'$, and it pierces the perspective plane at $d$, the perspective of $B,D'$. Hence $cd$ is the perspective of the line $AB—C'D'$, as seen from $EE'$.

*By vanishing points.*—$Ev—E'h'$ is the visual ray parallel to the given line, and hence meeting it only at infinity. It pierces the perspective plane at $h'$, which is therefore (190) the perspective of the point at infinity on $AB—C'D'$ produced. Again, $t''$ is the trace of the line $AB—C'D'$ itself; hence (202) $t''h'$ is the perspective of the indefinite line of which $AB—C'D'$ is a part, produced from $t''$ to infinity.

2°. *The oblique line.*—Let $AB—A'B'$ be the line. The operations being precisely similar to those of the previous case, $a'b'$ is the perspective of the limited line $AB—A'B'$, found by the visual rays from its extremities $AA'$ and $BB'$. Also, $v'$ being the trace on the perspective plane, of the visual ray $Ev—E'v'$ parallel to $AB—A'B'$, it is the vanishing point of the latter line, while $t'$ is its trace; hence $t'v'$ is the unlimited perspective of $AB—A'B'$.

*Having the projections of a line and its unlimited perspective*, its definite perspective can be found by means of either projection alone of the visual rays from its extremities. Thus

LINEAR PERSPECTIVE. 235

$a'$, perspective of $AA'$, is either the intersection of the projecting line $aa'$ with $t'v'$, or it is the intersection of $A'E'$ with $t'v'$.

In the same manner $c$ and $d$ could be found.

208. The fixed position of the two planes used in the last problem enables the reader to see the construction, especially of the vanishing points, unmixed with that of any transposition, as in (Probs. I.—IV.) of either of the planes. But the figure also illustrates the *inconvenience of the method by visual rays alone, when coupled with the use of two planes;* also that of *a fixed position of these planes.*

*First.* If the bisecting line of the horizontal visual angle (195) be, as is desirable, perpendicular to **V**, the line $EE'$ in Fig. 5, for example, would have fallen between $AA'$ and $BB'$, and the intersections of $A'E'$ and $B'E'$ with $aa'$ and $bb'$ would have been too acute for accuracy; and this difficulty would increase with the distance of the given point above or below the horizon $Hh'$.

*Second.* The fixed planes cause both projections and the perspective to be confounded together, a condition involving intolerable confusion of lines in finding the perspectives even of quite simple objects.

Seeing thus, experimentally, the inconveniences of Fig. 5, we will next illustrate the means of obviating them (200).

PROBLEM VI.—*To find the perspective of a rectangular prism one face of which is in* **H** *and another in* **V** *; and to find a point of its shadow on* **H**.

*In Space.*—In this problem will be illustrated—*1st*, the translation, forward, of the perspective plane; *2d*, the method of diagonals and perpendiculars (201, 203); *3d*, also that of vertical visual planes and auxiliary lines (204).

*In Projection.*—Pl. XXI., Fig. 6. *GL* is the first and *real* position of the ground line, **V** being also the perspective plane, and $G_1L_1$ is its second position; showing this plane to have been brought directly forward, parallel to itself, a distance equal to $GG_1$, before being revolved over into the paper, *i.e.* into **H**. $ABCQ$ is the plan of the prism, and $A'B'F'I'$ the elevation of its front face, which, being in the perspective plane, is its own perspective (202). $EE'$ is the point of sight, where $E'$, as usual, is shown only on the *translated* position of the perspective plane. $E'D'$, parallel to $G_1L_1$, is then the horizon (205).

*Diagonals and perpendiculars.*—Taking the upper, back, left-hand corner of the prism, whose projections are $C$ and $A'$, the horizontal projection of the *perpendicular* (203) from this point is $CA$ and its vertical projection and trace is $A'$. Then, $E'$ being (197) the vanishing point of perpendiculars, $A'E'$ is its perspective.

Again; $Ch$ is the horizontal, and $A'h'$ the vertical projection of the *diagonal* (203) from $C,A'$, hence $h'$ is its trace, and, drawing the diagonal visual ray $ED$—$E'D'$ parallel to $Ch$—$A'h'$, this ray meets the perspective plane at $D'$, the vanishing point (197) of diagonals; hence $h'D'$ is the perspective of the given diagonal, and by (201) $c$, where it meets the perspective $A'E'$ of the perpendicular, is the perspective of the corner $C,A'$ of the prism.

Likewise, by means of the ray of light $BB_1$—$B'B_1'$, we find the *projections* $B_1$ and $B_1'$ of the shadow of the corner $BB'$ upon **H**. Then $B_1'E'$ is the perspective of the perpendicular $B_1r$—$B_1'$ from this point of shadow, and $e'D'$ is that of the diagonal $B_1e$—$B_1'e'$ from the same point. Hence $b_1$, intersection of $B_1'E'$ and $e'D'$ is the perspective of the shadow $B_1$, and $I'b_1$ is therefore the perspective of the shadow of the vertical edge $B$—$I'B'$ on **H**.

*Visual planes.*—$QE$ is the horizontal, and $nn_1$ the vertical trace of the vertical visual plane containing the vertical edge at $Q$. Its vertical trace therefore contains $qk$, the perspective of that edge, which is limited at $k$ by the perpendicular $I'E'$, and at $q$, either by the perpendicular $B'E'$, or, since $c$ is already found, by $cq$, parallel to the ground line (192).

Likewise $c$ might have been found by the intersection of the perspective perpendicular $A'E'$, *or* diagonal $h'D'$ with the vertical trace $mc$ of the vertical visual plane $CE$.

Thus $F'A'cqkI'B'$ is the perspective of the given prism as seen from $EE'$.

Remembering that $GL$ indicates the real position of the perspective plane, $Et$, $= E'D'$ (207) is the real distance of the eye from the perspective plane; while $E't_1$ is its height above the horizontal plane.

EXAMPLES.—1°. Set the prism on end, with one face parallel to **V**.
Ex. 2°. Let its vertical faces be oblique to **V**.

THEOREM XIII.—*The vanishing points of rays and of their horizontal projections are on the same perpendicular to the ground line.*

A line and its horizontal projection are in the same vertical plane. Therefore, the same is true of the visual rays respectively parallel to them, these rays being both drawn through the same point, viz., the eye. Hence the traces of these visual rays on the perspective plane, both being in the trace upon the perspective plane of the vertical plane containing them, will both be in the same perpendicular to the ground line. Also the horizontal projection of any line, being horizontal, the visual ray parallel to it will be horizontal also, and hence the vanishing point of the horizontal projections of rays of light will be on the horizon.

PROBLEM VII.—*To find the perspective of a rectangular prism lying on* **H**, *but with its horizontal edges oblique to* **V**; *also its shadow on* **H**.

*In Space.*—The perspective plane shall be translated forward, as before, to avoid interference with the plan. But in addition, 1*st*, the auxiliary lines shall be those of the object (203, first); 2*d*, the shadow shall be found directly, and not from its projections; 3*d*, it will be shown how to avoid a difficulty which arises from inconveniently remote vanishing points.

*In Projection.*—Pl. XXI., Fig. 7. The first or real, and the translated ground line, coincide with those of the last figure produced. $EE'$ is the point of sight, and $E'V''$ the horizon, shown as in the last problem. $ABCD$ is the plan of the prism, and $m'p'$ the trace on the perspective plane of the plane of its upper base, which, by (203), makes a vertical projection unnecessary.

$Ev$—$E'v'$ is the visual ray parallel to the horizontal edges of which $AB$ is one, hence $v'$, the trace of this ray on **V**, is the vanishing point of all those edges. In like manner, by means of the visual ray $EV_1$,—$E'V_1'$ parallel to $AD$, we could find the vanishing point $V_1'$ (not shown) on the horizon, of $AD$ and of all parallels to it. But the triangle $vEV_1$ ($V_1$ the intersection of $EV_1$ with $GL$) is similar to $ABm$, $ADp$, or $pCm$. Hence, bisecting $Am$ at $m_1$, $Ap$ at $o$, and $mp$ at $n$, the lines $Bm_1$, $Do$, $Cn$ will be parallel, and the visual ray parallel to them will give their vanishing point $V''$ the same as if found by bisecting $v'V_1'$. Thus we have the necessary working points within the limits of the figure.

The vertical edge at $A$, being in the perspective plane, is its own perspective, as shown at $aa_1$, projected from $A$. Then $a_1v'$ is the indefinite perspective of the base edge $AB$, limited at $h$, perspective of the foot of the vertical edge at $B$, by $m_1''V''$, perspective of $Bm_1$ (in **H**) which thus replaces $m''V_1'$, perspective of $Bm$ (in **H**), $m_1''$ and $m''$ being the translated positions of $m_1$ and $m$, as shown by the lines $mm''$ and $m_1m_1''$, perpendicular to $GL$.

The upper edge at $CB$, and the line $Bm_1$, considered as in the plane of the top of the prism, pierce **V** at $m'$ and $m_1'$, hence $b$, perspective of the upper point $B$, is the intersection of $m'V_1'$ or of $m_1'V''$ with either the vertical $hb$ or with $av'$.

Likewise $q'$ and $p'$, being the traces of $CD$ considered respectively as in the lower and upper surfaces of the prism, and $o$ being translated to $o''$ and $o'$, and $n$ to $n'$; the point $r$, perspective of the lower corner at $D$, is the intersection of $q'v'$ perspective of $Cp$, with $o''V''$ perspective of $Do$.

Then $d$, perspective of the upper $D$, is the intersection of $p'v'$, with $o'V''$, or with $rd$, parallel to $aa_1$.

Finally $c$, perspective of $C$, is the intersection of $p'v'$ with $n'V''$.

*The Shadow.*—By (Theor. XIII.), and no particular direction of the light being given, assume $H'$ on the horizon as the vanishing point of horizontal projections of rays of light, and then $R$, making $H'R$ perpendicular to $GL$, as the vanishing point of the rays themselves. Then $bR$ is the perspective of the ray through the upper corner $B$ of the prism, and $hH'$ is the perspective of the horizontal projection of the same ray, hence $b_,$, their intersection, is the perspective of the shadow of $B$ ($b$) on **H**, and $hb$, is that of $bh$ on **H**. Then $b,V_1'$ will be the shadow, partly visible, of the upper edge $BC$.

Observing how previous vanishing points have been found (197), project $H'$ at $H$ then $HE$, and $RE'$ will be the projections of a visual ray parallel to the rays of light, and will thus show the direction of the light.

PROBLEM VIII.—*To find the perspective of a wall whose front face is sloping, and whose horizontal edges are oblique to* **V**; *also its shadow on* **H**.

*In Space.*—This problem will illustrate the method of diagonals and perpendiculars when **H** is revolved 180° (200), and the advantage of occasional recourse to other methods at certain points.

*In Projection.*—Pl. XXI., Fig. 8.  $GL$ is the ground line, $E'D$ the horizon, and $EE'$ the point of sight, since, when the horizontal plane is revolved 180°, the horizontal projection of this point must be as far behind the ground line as the eye really is in front of the perspective plane. Also, as in (Prob. VI.), $E'D = Et$.

$ABFI$ is the horizontal projection of the wall after the revolution of **H**. Then $ACBJ$, its sloping face, being the side towards $E$, is the real front of the wall. In fact, if, for example, $A$ and $B$ were as far behind the ground line, on the perpendiculars $Ao$ and $BL$ produced, as they now are in front of it, the line $AB$ would then be in its real position.

Finally, $d'r'$ is the vertical trace of the horizontal plane of the top of the wall.

So much done, the perspective of $B$, for example, in the base of the wall, is $b$, the intersection of $mD$, perspective of the diagonal $Bm$, with $LE'$, perspective of the perpendicular $BL$. The perspectives $a$ and $g$ of $A$ and $F$ in the base are similarly found, as shown.

The perspective, $c$, of $CC'$ in the upper base, is also similarly found, being careful to observe (203) that the diagonal $Ce$—$C'e'$, and perpendicular $Cn$—$C'$ from $CC'$, pierce the vertical or perspective plane at the height of the point $CC'$ whence they proceed—that is, at $e'$ and $C'$ in the trace $d'r'$ of the top of the wall.

The points $f$ and $j$, perspectives of $F$ and $J$ in the top of the wall, might be found in the same manner. But having $g$ already, $f$ may be found as the intersection of $gf$, perspective of the vertical edge at $F$, with $d'D$, perspective of the diagonal $FG$, or with the perspective (not shown) of the perpendicular at $F$. At $j$ the perspectives of the diagonal and perpendicular at $J$ would intersect quite acutely, and we use the vertical visual plane whose horizontal trace is $JE$, and whose vertical trace $hj$ intersects the diagonal $r'D$, for example, perspective of $Jr$, at $j$.

*The Shadow.*—Proceeding, as in the last problem, to bring all points within the limits of the figure, assume $H$, the vanishing point of horizontal projections of rays (205) and $R$ that of rays, and find $k$, perspective of the point $J$ in the base of the wall, by the intersection of the diagonal $rD$ and trace $hj$. Then $jR$ is the perspective of the ray from $J(j)$ and $kH$ is the perspective of its horizontal projection; hence $j_1$, their intersection, is the perspective of the shadow of $j$ on $H$; and $bj_1$ is thence the shadow of $bj$. The shadows $i_1$ of $I$ and of $f$ (not shown) may be similarly found, which will determine the direction of the small visible portion of the shadow of $FI$.

EXAMPLES.—1°. Let $E'$ and $D$ change places, and use the diagonals then appropriate.

Ex. 2°. Let $E'$ be above $d'r'$, and let the two projections of a ray of light be given.

PROBLEM IX.—*To construct the perspective of an interior containing various objects, by the method called that of scales.*

*In Space.*—This method, which most fully avoids the appearance of the *projections* of the given objects, depends mostly upon a principle, the simplest application of which is illustrated by a diagonal and a perpendicular. Thus a diagonal, as $DE$—$D'E'$, Fig. 6, cuts off, as we have seen, equal distances $Et = E'D'$ from a perpendicular, as $Et$—$E'$, and from a horizontal line $E'D'$ in the perspective plane. If, then, Fig. 6, we knew that the point $C,A'$ was at the height $F'A'$ above H, and at the distance $CA$ back of V, we could, for example, at once lay off, by a scale, $A'h'$, equal to the known distance $AC$ (numerically given), and draw the perspective diagonal $h'D'$ and perpendicular $A'E'$, *without drawing any plan, and hence without necessity for translating the perspective plane.*

The present problem illustrates this highly useful method, together with that of reduced vanishing points.

*In Projection.*—Pl. XXI., Fig. 9. $ABEF$ represents, on a scale of three feet to an inch, the section in the perspective plane of a room nine feet long and nine feet high. $CD$ is the horizon at the height of the eye (192), five feet from the floor. $C$ is the centre of the picture (207), five feet to the right of $AE$.

Supposing the room to be *seven and a half* feet deep, and that $Y$, one of the vanishing points of diagonals (not shown), is twelve feet to the right of $C$ on the horizon. Then the diagonal $q_1Y$ from $q_1$, $7\frac{1}{2}$ feet to the left of $B$, will cut off from $BC$, the perspective of the perpendicular at $B$, the perspective $Bb$ of $7\frac{1}{2}$ feet of depth into the room from the ground line $AB$.

But $Y$ being inaccessible, make $CD = \frac{1}{8}CY$, and correspondingly $Bq = \frac{1}{8}Bq_1$, and $qD$ will give the same point $b$ as before. For the triangles $CXY$ (where $X$ represents the eye itself 12 feet in front of $C$) and the triangle $Bq_1B_1$, whose perspective is $Bq_1b$, are similar, and with parallel bases $CY$ and $Bq_1$; hence (by Fig. 7, and as will at once appear on drawing the plan $ABB_1A_1$ of the room), dividing these bases proportionally, $CD = \frac{1}{8}CY$, and $Bq = \frac{1}{8}Bq_1$, $DX$, in space, will be parallel to the line $qB_1$, of which $qb$ is the perspective; hence $D$ is the vanishing point of this line (197).

Having $b$, the further side of the room is the square $bfea$, whose sides are limited by the perpendiculars $FC$, $EC$, and $AC$.

*The screen.*—Let this be 3 feet to the right of $A$; $1\frac{1}{2}$ feet into the room; $1\frac{1}{2}$ feet wide, and $2\frac{1}{2}$ feet high. Then draw the perpendiculars, $6C$ and $4\frac{1}{2}C$, and limit them at $H$ and $G$, by the parallel $gh$ to $AB$, from $g$, found by the line $vD$, drawn from a point $\frac{1}{2}$ foot from $B$. Make $AI_1 = 2\frac{1}{2}$ feet, and limit the perpendicular $I_1C$ at $k$ on the vertical $hk$, and draw the horizontal $kK$, meeting the verticals at $H$ and $G$, which completes the screen.

*The mirror and rod.*—Let there be a mirror 4 feet high and $1\frac{1}{2}$ feet wide on the left wall of the room, 3 feet from the floor, and $4\frac{1}{2}$ feet back of $AE$. Then on $AE$ make $A3$ and $A7$ at the heights of 3 feet and 7 feet, and the mirror will be included between $3C$ and $7C$. The construction can be completed as before.

The rod $ro$ parallel to $AB$ is assumed. Its true length would be from $m$ to the intersection of $Co$ produced with a parallel to $AB$ from $m$.

*The drop light and shadows.*—Let the light be $3\frac{1}{2}$ feet to the left of $BF$, 3 feet back of the perspective plane $ABEF$, and $3\frac{1}{2}$ feet long. Its intersection with the ceiling will then be the intersection $l$ of the perpendicular $3\frac{1}{2}C$ from the point $3\frac{1}{2}$ on $FE$, with the line $4\frac{1}{2}D$, or with $ty$, where $t$ is the intersection of $1D$ with $FC$, and $Ft$ is therefore the perspective of 3 feet from $F$ on $FC$. Then $L$, perspectively $3\frac{1}{2}$ feet below $l$, is the intersection of the vertical $lL$ with the horizontal $L_1L$, where $L_1$ is the intersection of the perpendicular $5\frac{1}{2}C$ with the vertical $tL_1$ on the right-hand wall.

Drawing now the vertical $yl_1$, intersecting $L_1L$ produced at $l_1$, and the parallel to $AB$ from $L_1$, intersecting $lL$ produced at $l_{\prime\prime}$, we have the projections, $l_{\prime}$ on the left-hand wall, and $l_{\prime}$ on the floor, of the light at $L$. This done, the shadow $k$ of $K$, upon the floor, is the intersection of the perspective ray $LK$ with its horizontal projection $l_{\prime}Gk$; and $i$, like shadow of $I$, is similarly found.

Also $s$, shadow of $o$ upon the left-hand wall, is the intersection of the perspective ray $Lo$ with its projection $l_{\prime}rs$.

209. *In extending this method of scales* to the case of objects oblique to **V**, as in Figs. 7 and 8, we may proceed in either of two ways.

*First.* We can ascertain, by scale, from temporary plans on waste paper, the necessary dimensions and distances estimated on parallels and perpendiculars to the perspective plane. Thus in Fig. 8, $Lo$ is the length of $AB$ estimated in the direction of the ground line, and $LB$ and $Ao$ are the perpendicular distances of $B$ and $A$ from the perspective plane, or their depth into the picture.

*Second.* See Fig. 7, drawing the arc $DD_1$ from $A$ as a centre, the chord $D_1D$, and any parallels to it, will cut off equal distances from $A$, both on $AD$ and on the ground line. This is the *general* principle of operation, of which the cutting off of equal distances, on the perpendicular and on the ground line, by a diagonal (as at $Ah = AC$, Fig. 6), is a particular case.

EXAMPLE.—Let a square be subdivided into any desired number of small squares, any of which at pleasure shall be, or shall circumscribe the bases of various geometrical solids, of various heights, and find the perspective of the whole. [See the next problem for cylinders and cones.]


PROBLEM X.—*To find the perspectives of two circles lying in* **H**, *the centre of one of them being, with the eye, in a profile plane.*

*In Space.*—In this problem, we shall use the method of diagonals and perpendiculars, aided by the principle that the perspectives of parallels to the ground line have, themselves, a like direction, and illustrating the convenience of circumscribing squares in finding the perspectives of circles.

The perspective plane is here translated forward.

*In Projection.*—Pl. XXI., Fig. 10. Let $PQRS$ and $P_1Q_1R_1S_1$ be the circumscribing squares of two equal circles, tangent to the ground line, which contains the sides $RS$ and $R_1S_1$ of the squares. $sr_1$ is the ground line of the translated position of **V**, the perspective plane, $E'D$ the horizon, $EE'$ the point of sight, and $D$ ($E'D = EK$) the vanishing point of diagonals.

The translated position $s$ of the trace $S$ of the perpendicular $PS$ and diagonal $QS$, is one point of their perspectives $sE'$ and $sD$, respectively. Then $q$, the intersection of $sD$ with $rE'$, the perspective of $QR$, is the perspective of $Q$, and enables us to draw $qp$, parallel to $sr$, for the perspective of $QP$. Likewise, $s_1$, $r_1$, $q_1$, $p_1$ are found, as shown.

Next, in both figures alike, as shown by like letters distinguished from each other by subscript numbers, draw the perpendiculars $BMN$, $CK$, $FJH$, whose perspectives $nE'$, $kE'$, $hE'$, meet the diagonal $sD$ at $m$, $o$, $f$, perspectives of $M$, $O$, $F$.

These points enable us to draw $mj$, $aoi$, $fb$, parallel to $sr_1$, giving $j$, perspective of $J$, on $hE'$; $a$ and $i$, perspectives of $A$ and $I$, on $sE'$ and $rE'$, and $b$, perspective of $B$, on $nE'$. Through the eight points thus found by means of only one diagonal, the ellipse $abfk$, perspective of the circle $ABFK$, can be drawn.

*Special points.*—We see that $co$ is less than $ok$. That is, the perspective $o$ of the centre $O$ is not the centre $v$ of the perspective; $v$ being the middle point of $ck$.

Draw the perspective diagonal $Dvg$, translate $g$ to $G$, and the diagonal from $G$ will give $V$, the point whose perspective is $v$, and $TVU$ the chord whose perspective is the transverse axis $tvu$ of the perspective ellipse. $T$ and $U$ can also be found as the points of contact of the horizontal traces, $ET$ and $EU$, of vertical visual planes. In either case, $t$ and $u$ can be found in advance, and thence the perspective ellipse constructed on its axes $ck$ and $tu$.

The axes of the other ellipse are less easily found.*

By precisely similar constructions, the perspectives of circles parallel to **H** and above it, or in planes perpendicular to the ground line, could be found. In fact, to realize the former case, it is only necessary to invert the figure, which will then, however, represent **H** as revolved 180°.

EXAMPLES.—1°. Construct Fig. 10 by the method of Fig. 8.

Ex. 2°. In Fig. 10 find the perspective of a circle in a *horizontal* plane whose vertical trace is anywhere above the horizon. [If equal to $ABIK$, and vertically over it, points of its perspective would be in vertical lines at $abcf$ . . . . . . $m$.]

Ex. 3°. Find the perspectives of two circles, one in a profile plane whose traces are $sS$ and $SP$, the other in the profile plane $r_1R_1Q_1$.

Ex. 4°. Find the perspective of a cylinder. [See Ex. 2 (note) and draw vertical common tangents to the upper and lower bases.]

Ex. 5°. Find the perspective of a cylinder whose axis is parallel to the ground line.

Ex. 6°. Find the perspective of a horizontal cylinder whose axis is oblique to **V**.

Ex. 7°. In Exs. 4, 5, 6, substitute a cone for the cylinder.†

### B—Perspectives of Developable Surfaces.

210. The apparent contour of the convex surface consists of the element of contact of tangent visual planes (D. G., Prob. XXX., etc.).

These planes are tangent to the visual cone whose base is a base of the surface, hence the perspectives of the elements of apparent contour are tangent to the perspective of either base.

Problems of this kind relating to single bodies are so easily solved by any of the methods already explained,‡ that we will take here but one, of a little more complication.

---

* See my Higher Perspective, Prob. LXV.
† See my Elementary Linear Perspective.
‡ See either my Elementary or Higher Perspective.

PROBLEM XI.—*To find the perspective of an elbow arch and of its shadows.*

*In Space.*—The elbow arch differs from the complete groined arch in that the two equal intersecting arched passages which form it, terminate at their meeting, instead of completely penetrating or crossing each other.

Thus, if Pl. XXII., Fig. 12, were the plan of a groined arch, composed of the semi-cylinders, one of which had the vertical semicircles projected on *JG* and *IK* for its bases, and the other those projected on *JK* and *GI* for its bases, the horizontal projection of the real portions of the former cylinder would be *JoG* and *KoI;* while the horizontal projection of the real portions of the other cylinder would be *JoK* and *GoI;* and the two cylinders would intersect in the two equal vertical semi-elliptical groins *JoI* and *KoG*.

But, taking the same figure as the plan of an elbow arch formed by the same cylinders, one groin as *KoG* would disappear, and the horizontal projection of their real portions would be *JGI* and *JKI*, where *GI* and *jl* would be elements of one cylinder, and *KI* and *Oo* would be those of the other. All the operations required in case of the groined arch will, therefore, be exhibited with less repetition, and on larger and less broken surfaces.

The problem will further illustrate the use of reduced vanishing points, lines in the direction *OI*, Fig. 12, being substituted for diagonals, as *JI*, and hence giving a vanishing point half as far as that of the diagonals from the centre of the picture; also the construction of shadows on cylinders parallel and perpendicular to **V**, and on profile planes.

*In Projection.*—1°. *The outlines.*—Let *AJKB* be the intersection of the perspective plane with the perpendicular passage at the beginning *J*, Fig. 12, of its intersection with the parallel one. This section is then seen in its real size on the scale of the drawing. *S* is the centre of the picture and *SD*, the horizon. $D_1$ is the vanishing point of lines parallel to *OI*, Fig. 12. *AS* and *BS* being then the perspectives of the perpendicular edges $JG$ and $KI$, Fig. 12, of the floor of the arch, they are limited at $D$ by $TD_1$, perspective of $OI$, and at $C$ by $DC$, parallel to $AB$.

The verticals at $C$ and $D$, equal in space to those at $A$ and $B$, are therefore limited at $G$ and $I$ by the perspective perpendiculars $ES$ and $FS$ (not shown).

Supposing $BN$ to be the thickness of the exterior walls, make $NU = \frac{1}{4}NA$ and $UD_1$ and $NS$ will intersect at the exterior corner $N_1$, whence the parallel $N_1 C$ to $DC$ can be drawn.

The side semicircle, $EhG$, can be found independently by lines parallel to $OG$, Fig. 12 (not drawn), but is here found from the groin.

2°. *The groin.*—Divide $OJ$, Figs. 11, 12, into four equal parts, when parallels to $OI$ from $L, Q, P$, will meet the groin at $l, o, p$. Now $Qo$ is in the level of $O$, the crown of the arch, but $Ll$ and $Pj_1$ are in the level of $Q_1$, Fig. 11, found by dropping the vertical $QQ_1$ to intersect the semicircle $EOF$, and drawing the horizontal $Q_1 J_1$.

This done, $o$, summit of the groin, is the intersection of the perpendicular $OS$ and $QD_1$, perspective of $Qo$, Fig. 12. Also, $l$ is the intersection of $Q_1 S$ and $LD_1$, and $p$ that of $Q_1 S$ and $PD_1$.

*Greatest apparent height of groin.*—This point is on the element of apparent contour of the parallel cylinder (210). To find it, suppose the structure to be revolved 90° to the left, for example, about a vertical axis through $o$. Then, it being square in plan, the vertical projections of the equal cylinders will replace each other, and $AJ$ will represent the perspective plane, and a point $S_2$ (not shown) at the distance of $2SD_1$ to the left of $S_1$, will represent the point of sight. The tangent $S_2 W$ to $EOF$ will then be the revolved trace of the visual plane tangent to the parallel cylinder, and $V$ a point of its trace on the perspective plane, which, by counter-revolution, appears at $Vw$ and is met by the perspective perpendicular $WS$ at $w$, the required apparent highest point.

*The lateral semicircle.*—This is here thus found from the

groin. $j$ and $j_1$ are the intersections of the perpendicular $\mathcal{J}_1S$ with the horizontal parallels $lj$ and $pj_1$, the latter confounded in the figure with $Vw$; and $h$, the apparent highest point, is the intersection of the perpendicular $W_1S$ with $Vw$.

3°. *The shadows.*—*That of EOF on the perpendicular cylinder and side wall* is found by planes of rays perpendicular to **V**, each of which will cut a point from the semicircle $EOF$, and an element receiving its shadow, from this cylinder. Thus, $SR$ being the vertical projection of the ray of light through the point of sight, assume $R$ as its vertical trace and consequently the vanishing point of rays, whence $H$ (Theor. XIII.) will be the vanishing point of horizontal projections of rays. Then $XY$, parallel to $SR$, is the trace on the perspective plane of a plane of rays, containing the element $YS$ of the cylinder, and the point $X$, whose shadow $x$ is the intersection of $YS$ with the perspective ray $XR$.

$Z$, where the shadow begins, is the contact of the tangent, parallel to $XY$.

Points of shadow of $EOF$ upon the side wall, as $x_1$, shadow of $X_1$, may be similarly found, as shown.

*Also thus:* $R_1TSO$, being a plane perpendicular to the ground line, $SR$ and $SR_1$ are the projections of a ray of light through $S$, and of its projection on the plane $R_1TO$. Hence $RR_1$ is the trace, on the perspective plane, of the plane of rays containing these lines, and consequently $R_1$ is the vanishing point of all lines parallel to the projection $SR_1$ of rays upon planes perpendicular to the ground line.

Then $X_1k$ is the trace on the perspective plane of a plane of rays through $X_1$, and perpendicular to $R_1TO$, hence the ray $X_1R$ meets its projection $kR_1$ on the plane $BK$, at $x_1$, perspective of the shadow of $X_1$ on $BK$.

*And again thus:* $X_1t$, being an ordinate of $EOF$, the shadow of $t$ on the side wall $Bk$ is $t_1$, intersection of $tR$ and $FR_1$, by the last method. Then $t_1x_1$, parallel to $tX_1$, is the shadow of $tX_1$, limited at $x_1$ by its intersection either with $X_1R$, $kR_1$, or $Y_1S$.

The graphical use of various methods is the advantage of each in cases where the adoption of others would give inaccurately acute intersection, as at the intersection of $t_1x$, with $kR_1$, or occasion confusion of new lines with previous ones.

4°. *The shadows on the parallel cylinder, rear wall, and floor.*—These are cast by both the front and side circles, and the vertical $AE$.

Drawing $AH$, the portion $AA_1$ is the shadow of $AE$ on the floor. At $A_1$ it rises, parallel to $AE$, on the rear wall, and is limited by the ray $ER$ at $e$, shadow of $E$; also found as follows.

The projection of $O_1$, centre of $EOF$, upon the rear wall is $o_1$, on $GI$. Then $o_2$ intersection of the ray $O_1R$ with its projection $o_1o_2$, on the rear wall, and parallel to $SR$, is the shadow of $O_1$ on that wall. Thence $o_2e$, parallel to $O_1E$ and limited by the ray $ER$, is the like shadow of $O_1E$, and radius of the circular shadow of $EO$, limited by $DI$.

A tangent from $R_1$ to the side circle is the perspective of the trace upon the plane $AJ$ of that circle, of a plane of rays tangent to the parallel cylinder, and hence gives $j$ (accidentally on $lj$) as the point (analogous to $Z$ on $EOF$), where the shadow on the parallel cylinder begins.

Secants from $R_1$, as $R_1sr$ and $R_1m$, are traces of similar but secant planes of rays, cutting points from the side circle, whose shadows fall on the horizontal elements as $su$ and $nq$, cut from the cylinder or the rear wall by these planes. Thus $u$, shadow of $r$, is the intersection of $su$ with the ray $rR$, and $q$, shadow of $m$, is the intersection of $nq$ with $mR$. The shadows $vqe$ and $x_1e$ are tangent to each other and to $A_1e$ at $e$, as the lines casting them are at $E$.

EXAMPLES.—1°. Replace the elbow arch by a complete groined arch, and make the figure fill the plate.

Ex. 2°. Find the perspective of two intersecting solid cylinders; one of them vertical, the other parallel to the ground line, with their shadows.

## Perspectives of Warped Surfaces.

211. There is seldom occasion to find these. That of the warped hyperboloid of revolution is best found in the way

to be explained for a similar double-curved surface. Screws and other practical examples of warped surfaces are always limited by simple and definite outlines, whose perspectives, easily found by methods already explained, will be, or will determine the perspectives of the given surfaces.

## Perspectives of Double-Curved Surfaces.

212. The method of three planes is in many cases preferable for problems under this head.

Also, in nearly all cases, an important part of the problem is the preliminary construction of the apparent contour, whose perspective is that of the given body.

213. The principle of (190) has important applications to the perspectives of double-curved surfaces. Thus, a sphere itself can never appear otherwise than round, or of circular outline, however viewed. Accordingly, if the perspective of a sphere be actually viewed from the real positions of the eye as given by its projections, or vertical projection and points of distance (207), this perspective will *appear* circular, though it may *really be* any one of the conic sections, and will so appear when viewed from other points than the actual point of sight. For the visual cone, whose vertex is the eye, and base its circle of contact with the sphere, will be a cone of revolution. Now the perspective plane may be taken so as to cut this cone in any one of the conic sections. But the section of the visual cone by the perspective plane is the perspective of the given sphere, which may thus be any one of the conic sections.

In fact, the perspective of a sphere will only *be* as well as *appear* circular, when the visual ray to its centre is perpendicular to the perspective plane.

EXAMPLES.—1°. Find the perspectives of four spheres, the centre of one of them being in the perpendicular from the eye to the perspective plane, that of another being at any other point in the plane of the horizon, that of the third being below, and that of the fourth above this plane. [By the method of either two or of three planes, all secant vertical visual planes (204) will cut the spheres in circles, great or small, tangent visual rays to which, easily shown on H or V in revolved position, will give points of apparent contour and their perspectives.]

Ex. 2°. Find one or more points of the curve of shade on each sphere. [The most elementary method being to find first their projections.]

Ex. 3°. Find the perspective of a hollow hemisphere with its *visible* interior and shadows, limited by a great circle parallel to V.

Ex. 4°. Find the perspective of a hollow hemisphere with its *visible* interior and shadows, limited by a great circle parallel to H.

PROBLEM XII.—*To find the perspective of the concave double-curved surface of revolution, called a piedouche or scotia.*

*In Space.*—This problem will illustrate the method of three planes, as applied to double-curved surfaces; together with the preliminary construction for finding the *projections* of the contour and the shade. To aid in becoming familiar with the method, the eye is, for variety, placed to the left of the object, and the perspective plane translated about its vertical centre line instead of about its vertical trace.

*In Projection.*—Pl. XXII., Fig. 13. The given body, $ADBC—A'B'C'$, is called a *scotia*, when of large diameter compared with its height, as when found in the base of a column, and a piedouche, or pedestal, when proportioned as in the figure. The meridian curve $A'I'q'j'$ may be compounded of unequal quadrants, or, as in the figure, may be a semi-ellipse on $A'j'$ as a diameter.

$D_1A_2$ is the real, and $d_1a_2$ the translated position of the perspective plane, which is revolved about the vertical at $O$ to the position $d_2e_2$, parallel to V. The place of the eye is $EE'$ (at the left of Fig. 11), and $GL$ is the ground line.

1°. *The projections of the apparent contour.*—This is the contour seen from $EE'$. Two points of it are $UU'$ and $H$ ($H'$ not lettered) the points of contact of the vertical visual planes represented by tangents from $E$, to the smallest circle of the concave surface. Two others are the points of contact, near $Fj'$ and $BB'$, of visual planes perpendicular to $V$, and hence shown by tangents from $E'$ to the meridian $Aq'j'$ and $B'p'$.

Intermediate points are found by auxiliary tangent cones or spheres just as in (shadows, Prob. XIII.), only that the rays considered converge to $EE'$ instead of being parallel.

Thus, assuming any parallel $p'q'$—$pq$ (33) of the surface, the tangent $p'Q'$ at $pp'$ meets the axis $Q'C'$ of the surface at the vertex $Q'$ of the tangent cone having $p'q'$ for its circle of contact. Then the visual ray $Q'E'$—$CE$ meets the plane $p'q'$ of that circle at $R'R$, whence the tangents $Rc$ and $Rc_1$ give the two points $cc'$ and $c_1c_1'$ of the required contour. The points $ll'$ and $l_1$ are similarly found.

*But note*, that only so much of this contour is *real* on an opaque solid of this form as is between the points as $MM'$ on the lower part, and $KK'$ on the upper part of the surface, where visual rays are tangent, not only to the surface of the body, but to the curve of contour itself; because, for example, tangents to the surface between $M$ and $T$, on the right, are also secants, having first pierced the surface somewhere to the left of the chord $MT$.

2°. *The curve of shade.*—Recalling the principle of (178) let $R_1C$—$R_1'Q'$ be the projections of a ray of light. The curve of shade is then found by constructions precisely similar to those for obtaining the contour. Thus, for intermediate points, for example, taking the same cone $p'q'Q'$ as before, the ray of light $CR_1$—$Q'R_1'$ pierces the plane of its base at $R_1'R$, so that the tangents from $R$ to the base $pq$, being the traces, on the plane of the base, of tangent planes of rays, give $ss'$ and $s_1s_1'$ two points of shade.

3°. *The perspective.*—This is now found essentially as in Prob. I. Thus, the visual ray $DE$—$D'E'$ pierces the perspective plane at $D_1D_1'$, which is thence translated on $D_1d_1$—$D_1'd$ parallel to $a_2L$, to $d_1$, and then revolved to $d_2$ and projected at $d$, the perspective of $DD'$.

Likewise $m$, extremity of the visible contour of the concave surface, is the perspective of $MM'$, and all other points, though not fully shown, of the contour and shade are found in the same way.

4°. *Method by perspectives of parallels.*—Sometimes, as a substitute for the preliminary construction of the *projections of the contour,* when this is not familiar, we may find by Prob. X. the perspectives of numerous horizontal circular sections of the given body, and then the perspective of the visible contour will be a curve which can be sketched tangent to the perspectives of all these circles.

EXAMPLES.—1°. Filling a plate with the perspective only of the piedouche [the projections being on adjacent temporary sheets], find the perspective of its curve of shade, and of the shadow of the upper cylinder [fillet] upon the concave surface.

Ex. 2°. Find the perspective of an ellipsoid, a vase, an annular torus, or a bell, by the method (4°) just indicated.*

### Adhemar's General Method.

214. Viewing perspective as a direct application of descriptive geometry, the method of three planes (196) may be considered as its most general method.

But in the present case, perspective is considered rather in relation to *practical convenience of construction, combined with the best pictorial effects, obtained without constraint as to the relative position of the object and the observer.*

The method of Adhemar is universal in fulfilling all of these conditions; so that under the first no point of construction shall ever fall without the convenient limits of the paper.†

---

* Referring the reader to various accessible works, among them both my Elementary and Higher Perspectives, for abundant perspective examples as usually treated, we will, by preference, next give what may not easily be found elsewhere, a fuller account of the practical method of scales by ADHEMAR.

† "Before publishing this method in the first edition of my work, I had wished to assure myself that it was always applicable, and to attain this object I imposed upon myself the condition of executing all the plates of my atlas [81 (folio) in number] *without ever employing a single point without the plate.*" Adhemar, 3d Ed., 1860.

In relation to its principal determining lines, this method is called that of *dimension and vanishing scales*, and has been partially illustrated in Prob. IX.

**PROBLEM XIII.**—*To find the perspective of a point taken at pleasure in* **H**.

*In Space.*—Pl. XXII., Fig. 14, represents an auxiliary sheet containing a plan view, on a smaller scale, and used where it is desired to make a large perspective view. $EP$ and $EQ$ represent the extreme visual rays, chosen so as to include a suitable visual angle (195) generally not varying materially from 45°. $Ec$, the central visual ray, bisects the visual angle, and the perspective plane $PQ$ is perpendicular to it. $PS$ is the horizontal trace of a plane perpendicular to the trace and plane $PQ$, and $A$ is a point in $H$ at a perpendicular distance $AB$ from $PS$.

Regarding $EP$ and $EQ$ as horizontal traces of vertical visual planes, the two vertical edges of the picture, at $P$ and $Q$, are their traces on the perspective plane.

*In Projection.*—Passing now to Fig. 15, $PQ = 2\frac{1}{2}PQ$ of Fig. 14; $PQRS$ is the perspective plane standing perpendicular to the paper at $PQ$, Fig. 14, $VC$ is the horizon, and $C$ the centre of the picture.

Then take $Cx = cP$, Fig. 14, and drop $xp$, perpendicular to $VC$, and meeting $PC$, perspective of $PS$, Fig. 14, at $p$. Make $pq = AB$, Fig. 14, and draw $Cq$. Also make $CV = Ec$, Fig. 14, and $PB = PB$, Fig. 14, and draw $BV$, and from $b$, its intersection with $PC$, draw $ba$, parallel to $PQ$. Then $a$, intersection of $ba$ with $Cq$ produced, is the perspective of $A$, Fig. 14, $Cq$ being the perspective of the perpendicular from $A$.

*Demonstration.*—Produce $Cq$ to $A$ on $PQ$, Fig. 15.

Then $PA : pq :: PC : pC :: RC : xC (= Pc,$ Fig. 14$) :: \frac{1}{2}PQ : \frac{1}{2}PQ$ (Fig. 14) $:: PQ : PQ$ (Fig. 14),

whence, finally,

$$PA : pq :: PQ : PQ \text{ (Fig. 14)}.$$

LINEAR PERSPECTIVE.      255

Thus, by laying off $AB$ once on $pq$, we obtain the same result as if we had made $PA = \frac{4}{5}AB$ (Fig. 14), and $PB$ and $CV$ each $\frac{4}{5}$ of their present size.

It may be objected to this method that it determines longer lines, as $ba$, by shorter ones, as $pq$; but in its applications, where numerous lines are to be found, it will be seen that, by means of numerous mutual checks, points are very accurately found.

PROBLEM XIV.—*To find the perspective of any quadrilateral in* **H**.

Pl. XXII., Fig. 16, represents a temporary sheet containing the plan $ABCD$ of the quadrilateral, the perspective plane at $PQ$ and the visual angle $PEQ$ ($E$, the intersection of $kP$ and $cQ$) all on the same convenient scale. $Ps$ and $Qc_1$ are perpendiculars to $PQ$ at $P$ and $Q$.

Pl. XXII., Fig. 17, represents the perspective plane $PQVV_1$, where $PQ = 3PQ$, Fig. 16, the horizon $VV_1$, centre of the picture $E$, *vanishing scales* $PE$ and $QE$, perspectives of $Ps$ and $Qc_1$; and $pq$, scale of *dimensions*, all found as in Prob. XIII., Fig. 15. That is: $EV_1 = EV_1 = rE$, Fig. 16, and similarly always; and $Ex = Pr$, Fig. 16.

So much done, the perspectives of the corners, $A, B, C, D$, of the given quadrangle, may be found by ordinates to $Ps$ or $Qc_1$, either from those corners, or from the intersections of the sides of the quadrangle with the visual rays $Pk$ and $Qc$, or with $PQ$. The latter is more accurate, as determining required points *between* points of construction.

Proceeding as in Prob. XIII., make $PF' = PF$, Fig. 16, and draw $F'V$, and $F$ will be the perspective of $F$, Fig. 16. In like manner the intersections of all the sides with $PE$ ($Ps$) or $QE$ ($Qc_1$) could be found, as shown at $h$.

But produce the sides to give $n, a, b, c$, etc., Fig. 16. Then, for example, make $Qc' = Qc_1$, Fig. 16, and draw $c'V_1$, and thence $c_1c$ parallel to $PQ$, and $c$ is the perspective of $c$, Fig. 16. Also find $a$ and $b$ similarly.

256                    LINEAR PERSPECTIVE.

Again, for example, make $pn_1$ and $pd$ respectively equal to $Pn$ and $Dd$, Fig. 16, and $En_1$ and $Ed$ produced, will respectively give $n$ and verify $D$. Now draw $aF$ and $cn$, which give $A$. $N$ and $S$ are found by making $Pe'$ and $Ps'$ respectively equal to $Pe$ and $Ps$, Fig. 16, drawing $e'V$ and $s'V$ to meet $PE$ at $e$ and $s$, perspectives of $e$ and $s$, Fig. 16, and thence $eN$ and $sS$, parallel to $PQ$, and intersecting $Er$ at $N$ and $S$, perspectives of the same points on Fig. 16. Then $Sb$ and $nc$ give $B$, etc.

We have thus, first, accurately made a perspective figure much larger than the usual methods would produce from the given plan; and, second, have done it without the need of any point of construction outside of the picture.

PROBLEM XV.—*To find the perspective of a stone bridge.*

*In Space.*—This concluding problem will illustrate the following points.

1°. The compact but accurate construction, by means of suitable checks, additional to those before given, of large perspective plans from small projection plans.

2°. The construction of perspective altitudes.

3°. The construction of perspective elevations by throwing them up from auxiliary perspective plans; a method which is particularly useful in the perspective drawing of more complicated structures.

4°. The independence of the plan and perspective in respect to scales, and thus the entire generality of the method.

*In Projection.*—Pl. XXIII., Fig. 5, represents the required perspective figure, while Figs. (A), (B), (C), (D), are supposed to be, in practice, on temporary sheets fastened adjacent to the drawing-paper, Fig. 18, though Fig. (C) or (C) and (D) may at pleasure be on the same sheet with Fig. 18, and Fig. (C) may be either above or below Fig. 18, as most convenient.

In Fig. (A), the visual angle $AEB$ includes two piers and a part of a third one. $AB$ is the perspective plane, $E$ the point of sight, $v$ the vanishing point of lines parallel to the sides, as $ab$, of the piers, and $Ds'$ the trace of a vertical plane parallel to the visual plane $EB$; also $EA$ happens to coincide with the face $si$ of a pier.

The *lines* of construction relied upon are the lines, as $sf$, containing the points of the piers, the traces of the planes, as $hg$, containing the longer edges of the piers or the axes, as $rr_1$, of the arches, and parallels, to $AB$, as $ss'$ or $JJ_1$, from corners of piers, or from traces of the previous lines upon the visual planes or upon $Ds'$.

The *points* used in determining the perspectives of these lines are their traces, on the limiting visual planes, or upon $AA_1$ (Fig. C), perpendicular to $AB$, or on $Ds'$, or $AB$.

Fig. (B) is a partial side and end elevation on the same scale as Fig. A.

1°. *The perspective plan*, Fig. (C). This is found as follows: Having made $Cx = \frac{1}{2}AB$, Fig. (A) (the letters in parentheses denoting figures), and drawn $xp$ and $pP$, and taken $V$ as in the two previous problems, and $AB$ at pleasure; make $AJ_2 = AJ_1$ (A), and draw $J_2V$, giving $J_1$ on $AC$, perspective of $J_1$ on $AC$ (A). Then making $qP = BD$ (A) or $BD$ a fourth proportional to $AB$ (A), $AB$, $BD$ (A), we have $DB_1$, the perspective of $Ds'$ (A), since $B_1$ on the horizon $VCB_1$ is the vanishing point of all parallels to $EB$ (A). Now draw $J_1J$, perspective of $J_1J$ (A). Next make $Ak_2 = Ak$ (A), draw $k_2V$, giving $k$ on $AC$, and $kK$, parallel to $AB$, giving $K$; then $KJ$ is the perspective of $KJ$ (A). Finding $Kk''$ and $ii''$, similarly, perspectives of $Kb$ and $ic$ (produced to meet $Ds'$ produced), $KK_1$, perspective of $KK_1$ (A), is accurately found by the intersection of $KJ$ with $Kk''$ and $ii''$. Also $K_1$ can be directly found by an ordinate parallel to $Kk$.

All other points of the perspective plan, on $AA_1$, $AC$, $BB_1$, or $DB_1$, can be similarly found by means of the *lines* and *points* of construction already stated.

2°. *The perspective elevation*, Fig. 18. Project the points as $s$, $i$, etc., of $AA$, upon $DB_1$ by parallels to $AB$, as at $s'$, $i'$, etc., and then erect the verticals $s'f_1$, $i''U$, $PH$, etc., parallel to $BB_1$. Then, taking $cV_1$ as the horizon, $D_1c$ becomes the new perspective of $Ds'$ (A).

At $P'$, intersection of $PH$ with $D_1c$, make $P'U' = uu''$ (B). Then, on $P'U'$ lay off from $P'$, the heights of the several characteristic horizontal lines of the bridge above $ss'$ (B)—that is, the heights of the horizontals through $f, o, e, y$ (B). Through the points on $P'U'$, thus found, draw lines, as $cP'P_1$, from $c$, giving the heights $f_1f_2$, on the vertical at $s'$, equal to that at $s$, its primitive position, and $O, E, Y, U$, above $P_1$ on the vertical at $i'$, equal to that at $i$ (A).

The corresponding points on the verticals at $s''$, $i'''$, etc., are their intersections with the same radials from $c$. This rather inaccurate determination of larger distances from smaller ones may be improved by making $BD$ (A) larger, and checked by making $P_1U$, for example, a fourth proportional to $cP'$, $P'U'$, and $cP_1$, by the usual construction.

Having thus found the points on these verticals, project those on $s'f_2$, $i'U$, etc., at $o', e', y', u'$, etc., on $A_1u'$, and join these points with $o_1, e_1, y_1, u_1$, etc. Thus $u'u_1$ is the *front* top edge of the bridge. The *back* top edge would be drawn from the projection of $U_1$ upon $A_1u'$ to the intersection of $cU$ with the vertical at $k''$.

3°. *The arch points* are thus found. Describe an arc with radius $OY$ and draw the tangent $RT$ at 45° with $TO$, and the radius $OR$, and ordinate $RN$, giving the new points $T$ and $N$, and thence the lines $t't_1$ and $n'n_1$, as before. Then projecting $o''y''$, etc., from $o, g$, etc., on the perspective plan, we have the perspectives, as $o'y'y''o''$, of the circumscribing squares of the front semicircles of the arches, and, by projecting $K_1$ and $r_1$ on $o'o_1$ and $t't_1$, we find the perspectives of their centres, and of the points projected at $T$; also $oy'$ and $oy''$, the perspectives of the diagonals of these squares, intersect $n'n$, at $n$ and $n''$, the perspectives of the points of contact of the tangents $tn$ and $tn''$.

Having thus five points, as $o', n, y, n'', o''$, of each arch-curve, and the tangent at each, the curve is easily sketched. The visible portions of the back arch-curves are similarly

# LINEAR PERSPECTIVE.

found from like points derived from points on the verticals at $K'$ and $k''$, and from points on $Kk''$ (C).

$V$, found by making $VA_1$ a fourth proportional to $AB$ and $Av$ (A) and $AB$, is the vanishing point of plan perspectives of $ah$, etc. (C), and $V_1$, projection of $V$ on $cV_1$, is that of the perspective of $ah$, etc., on Fig. 18.

Finally, we see, from the necessity of Figs. (A) and (B), the dependence of perspectives on projections, and thus here, as well as in the figures of Plate XX., that perspective is an application of descriptive geometry (194).

# SPHERICAL PROJECTIONS.

**215. Spherical projections** are the projections of such and so many of the circles of the earth, considered as a perfect sphere, as will permit the representation of any given points of its surface, and of the zones which are marked by the limiting apparent positions of the sun.

These projections are commonly made upon the plane of some great circle, but sometimes upon a tangent plane, and sometimes upon one or more circumscribing tangent cones, which, by development, will also afford plane representations.

216. The necessary circles are: *first*, certain *great* circles. Of these are:

*The equator*, whose plane is perpendicular to the axis of the earth.

*The ecliptic*, the intersection of the earth's surface with a plane containing its centre and that of the sun, and making an angle of very nearly $23\frac{1}{2}°$ with the plane of the equator.

*Meridians*, which are all in planes containing the axis, and hence perpendicular to the equator.

217. There are also the following *small circles*, all in planes parallel to that of the equator:

*Parallels*, taken at any convenient constant number of degrees apart.

# SPHERICAL PROJECTIONS. 261

The *tropics*, Cancer and Capricorn, tangent to the ecliptic, and therefore $23\frac{1}{2}°$ distant from the equator, and limiting the central or torrid zone, beyond which the sun's rays are never vertical.

The *polar circles*, arctic and antarctic, $23\frac{1}{2}°$ from the north and south poles, respectively, and whose distance, on a meridian, from the remotest point of the ecliptic, is 90°. These mark the frigid zones, outside of which the sun's greatest apparent altitude is never less than $23\frac{1}{2}°$.

218. The point of sight may be at an *infinite* or at a *finite* distance.

In the former case, the projection is always made upon a plane, is called *orthographic*, and can easily be constructed by any one familiar with orthographic or perpendicular projections (5—6).

The plane of projection is, in spherical projection, often called the primitive plane, and its intersection, or contact, with the sphere, the primitive circle, or point, respectively.

219. *The orthographic projection* is usually made either upon a plane containing the axis, as that of a meridian, or perpendicular to it, as that of the equator.

*In the former case*, the projection of every *parallel* would be a straight line perpendicular to the axis, and of every *meridian*, an ellipse, whose transverse axis would be the axis of the sphere and whose conjugate axis is the projection of the diameter perpendicular to the axis, upon the plane of projection.

Thus, Pl. XXIII., Fig. 19. *NESW* is the meridian in the plane of the paper, *NS* the axis, and the meridian perpendicular to the paper, and *NdS* the meridian at an angular distance equal to *Nc'* to the west of *O;* recollecting that the circles *NESW* and *EOW* are equal and perpendicular to each other, and that *NWS* may be considered as *NdS* after revolution about *NS* into the plane *NESW*, so that by (Prob. XXI., 2°) $ON(= Oc')$ : $Od$ :: $nc'$ : $nc$.

220. In the orthographic projection of the sphere on the plane of the *equator*, the *parallels* would be circles concentric with the equator, and the *meridians*, diameters of the circle representing the equator.

*The advantage* of this method to all acquainted with descriptive geometry, is its great theoretical simplicity.

*Its disadvantages* are the practical difficulty of constructing the elliptical meridians and the more serious one of the great distortion, by crowding, the nearer we approach the great circle which is in the plane of projection.

221. *The eye at a finite distance* (218). All cases under this head are really *perspectives* of the sphere (189) varying with the positions chosen for the eye, and the kind and position (215) of the surface on which the projection is made, all of which are quite numerous.

222. *Stereographic projection.*—In this case, the eye is at the pole of the primitive circle, as at the north or the south pole, if the projection be made on the plane of the equator. Pl. XXIII., Fig. 20. By this projection, only the hemisphere beyond the primitive circle is projected from the same point of sight, and it is mechanically easy of execution, since it possesses the curious property that every circle of the sphere not in a plane through the eye is projected in a circle, since, as will be immediately evident by a figure, the section cut by the primitive plane from the visual cone (194) whose base is any circle of the sphere, is a sub-contrary section of that cone (96). The distortion by this method is, however, still considerable, and is greater *by diminution of true size* near the centre of the primitive circle, for $E$ being the point of sight, $NOS$ the primitive (218) or perspective plane, and $NWS$ the hemisphere projected from $E$, the perspective $a_1b_1$ of $ab$ is less in proportion to $ab$ than $c_1d_1$, perspective of $cd$, is in proportion to $cd$.

In other words, $be$ being a semi-parallel, and hence $e$, in the perspective plane $NS$, being its own perspective, $b_1$, the perspective of $b$, is further below $eb$ than $d_1$ is below $d$.

# SPHERICAL PROJECTIONS. 263

**223. Construction.**—From the ease with which the stereographic projection is made on the plane of a meridian, we give the construction.

Let $NS$, Fig. 21, be the axis, $EW$ the equator, $O$, shown after revolution 90° about $NS$ at $E$, the point of sight, and $NESW$ the given meridian. Then $ab$ being the vertical projection of a parallel, $a$ and $b$ are their own perspectives as seen from $O$, and $f$ is that of $a_1$ as shown by the revolved construction $aE$. Then the centre $d$ of the *parallel afb* is the intersection of the tangent at $a$ with the axis $SN$.

For call $g$ the middle point of $af$. Then the triangles $dgf$ and $EaW$ are evidently similar, hence their doubles $daf$ and $Eah$ are so also, $Wh$ being made equal to $Wa$. But the angle $aEh$ is measured by $Wa$, which therefore measures the equal angle $adf$. But $Wa = \frac{1}{2}(SWa - Na)$ which last measure shows $da$ to be a tangent.

Again, to draw any *meridian*, as the one 30° distant from $W$, draw $Sh$ making $OSh = 30°$, then $h$ is the centre and $hS$ the radius of the required meridian $NhS$. This will be obvious on making a horizontal projection, when $O$ will appear as at $S$, and the horizontal projections of meridians will be straight lines through the centre.

**224. Globular projection.**—Pl. XXIII., Fig. 22. In this, the eye is exterior to the sphere, and at a distance $Sd$ from it, on the axis $NS$ of the primitive circle $EW$ equal to the sine of 45° = $\sqrt{\frac{1}{2}}$, the radius of the sphere being unity.

Result: $Nb$ being 45°, $ab = \sqrt{\frac{1}{2}}$, $aO = \sqrt{\frac{1}{2}}$; $OS = 1$; $Sd = \sqrt{\frac{1}{2}}$

but $\quad\quad\quad da : ab :: dO : Oc$

or $\quad\quad\quad 1 + 2\sqrt{\frac{1}{2}} : \sqrt{\frac{1}{2}} :: 1 + \sqrt{\frac{1}{2}} : Oc$

$\therefore \quad\quad\quad Oc = \dfrac{\frac{1}{2} + \sqrt{\frac{1}{2}}}{1 + 2\sqrt{\frac{1}{2}}} = \frac{1}{2} = Ec.$

That is, the globular projections $Oc$ and $Ec$ of the equal arcs $Eb$ and $Nb$ are equal, and the like is nearly true of other equal divisions of $NbE$.

But, on the other hand, the projections of circles not parallel to the primitive plane will generally be ellipses.

**225.** *Approximate globular, called equidistant, projection.* In this, which is often called globular projection, and supposing it made on the plane of a meridian, divide $EW$ and $NS$ into *equal* parts, also $NbE$ and $NW$ into the same number of equal parts as $NO$.

This done, *meridians* will be represented by arcs, each of which is determined by three points, viz., the poles, and one of the points of equal division on the equator $EW$. The *parallels* will also each be determined by three points, viz., a point of division on $ON$, and the points similarly numbered on $NE$ and $NW$, reckoning from $O$, $E$ and $W$.

**226.** The three foregoing projections are the only ones which are adapted for representing the entire surface of the earth.

They can be sufficiently well compared by making maps of the northern half of the western hemisphere, on the same scale for each, on the given meridian $NESW$ in the plane of the paper. The comparative amount and place of distortion will then be apparent.

**227.** Remembering that a sphere is undevelopable, no map of the world can represent the surface of the earth on a uniform scale as in plotting a survey of a tract sufficiently small to be regarded as plane.

The following methods, of more or less approximately uniform scale, are applicable to greater or less portions of the earth's surface.

**228.** *Cylindrical projection*, Pl. XXIII., Fig. 23. Here the projection is made upon a circumscribed tangent cylinder $WEMK$, the eye being at the centre of the sphere. Any parallel as $cd$ is thence projected by a visual cone whose vertex is $O$, in the circle $CD$, intersection of this cone with the cylinder. The meridians are projected upon the cylinder by their own planes, in parallel elements of the cylinder. The cylinder is then supposed to be developed, when meridians and parallels will both appear as parallel straight lines, each crossing the other at right angles.

It is obvious that this method would not apply within a certain distance of the poles, owing to the extreme distance of the projections of the parallels from the equator. Indeed, this projection is but very little used.

229. *Mercator's projection.*—This is a modification of the pure cylindrical projection, and is almost exclusively used for mariners' charts, since it is so constructed that the path of a vessel sailing so as to cross the meridians at a constant angle will be projected in a straight line.

It is very easily shown that the length of a parallel or of any part of it, $ab$ or $ad_1$, Fig. 19, is equal to the great circle $WE$, or arc $Wd$, respectively, multiplied by the cosine of the latitude $Wa$ of that parallel. Hence, if, instead of making the meridians converge according to this law, they are drawn as equidistant parallel straight lines, the parallels, instead of being equidistant as on the sphere, must be drawn at distances apart which shall increase according to the same law. That is, arc $Wd = \dfrac{\text{arc } ad_1}{\text{cos. lat.}} = \text{arc } ad_1 \times \text{sec. lat.}$, or the distances between the parallels must increase as the secant of the latitude increases.

In the accurate construction of such charts, tables giving the increments of the secant for each minute of latitude, usually up to 80°, are prepared.

230. *An approximate graphical construction*, sufficient for use in gaining an idea of the appearance of a map on Mercator's projection, is as follows: Supposing the quadrant $NE$, Fig. 23, to be on a large scale, $OE = 1$ foot, for example, divide $OE$ into equal parts—the more, the greater the accuracy—for example, 30—and number them from $O$, which will be marked zero, and draw the meridians parallel to $ON$,

through the points of division. Divide the quadrant $EN$ into the same number [30] of equal parts, therefore of 3° each, and number them from $E$ as the zero point, and draw radii through the points of division. At 1 on $OE$, the first point of division to the right of $O$, draw a parallel to $ON$ and note its intersections, $a$ with radius $O1$, $b$ with radius $O2$, $c$ with radius $O3$, etc. Then $Oa$, laid off from $O$ on $ON$, will give $a'$, the point through which the parallel of 3° will be drawn, like all the parallels, parallel to $OE$. Likewise $Ob$, laid off from $a'$, will give $b'$ on $ON$, a point of the parallel of 6°; $Oc$, laid off from $b'$, will give $c'$, a point of the parallel of 9°, etc.

All the parallels and meridians thus located need not be actually drawn; every fifth meridian will be enough. These being 15° apart, are called hour-circles, as they divide the circumference into 24 equal parts.

Parallels are usually 10° apart. To make them so, divide the quadrant into 18 or 36 equal parts for example, of 5° or 2½° each, and replace the line $1abc$ by a parallel line $a_1 b_1 c_1$, distant from $ON$ by $\frac{1}{18}$ or $\frac{1}{36}$ of $OE$, as the case may be.

231. *Gnomonic projection*, Pl. XXIII., Fig. 23. In this, the eye is at the centre of the sphere, which is then projected upon a tangent plane; most conveniently that at one of the poles, as $BN$. The meridians will then be projected in straight lines, making the same angles with each other at the point of contact $N$, of the plane, that the meridians themselves do. The parallels, as at $b$, will be projected in concentric circles around the same point, and whose radii, as $NB$, are equal to the tangents of the distance $Nb$ of the parallel from the pole, called the *polar distance*.

232. *A modification of this projection*, called *polar projection*, consists in taking the plane of projection through the polar circle. The meridians will be projected as in the last case, but the parallels will be projected in smaller circles than before.

Both of these projections are used to supplement Mercator's projection in representing the polar regions, to which that will not apply (228—9).

**233. *Conic projection.*** —In this projection, a cone tangent along some parallel is substituted for a tangent cylinder, the point of sight still remaining at the centre of the sphere. On developing the cone to form the map, the meridians will appear as elements of the cone, and the parallels as arcs having the development of the vertex for their centre.

This method is quite inaccurate for parallels far from the parallel of contact, and inconvenient for parallels of contact near the equator, on account of the remoteness of the vertex, in that case.

To increase the average accuracy within certain limits, *a secant cone*, containing two parallels not far apart, is sometimes substituted for a tangent cone.

**234. *Polyconic projection.*** —In this, a separate tangent cone is used for each of as many parallels as may be desired. Each of these parallels then appears in the development as an arc whose radius is the tangent of the polar distance of that parallel (231).

The central meridian of the map is made straight, and equal to the *true length* of the arc of that meridian contained within the limits of the map. The scale is therefore exact and uniform on this meridian and on the developed parallels; which intersect this meridian at their true distance apart.

The other meridians are curved, crossing the parallels at right angles, and are tangents at those points to the developed elements of the respective cones whose bases are these parallels.

See Figs. 1 and 2 in the text. In Fig. 1, $NS$ is the axis of the sphere, and $ad$, $bc$, $cf$, three parallels, curves of contact of the three tangent cones whose vertices are $V$, $V_1$, $V_2$, respectively.

268    SPHERICAL PROJECTIONS.

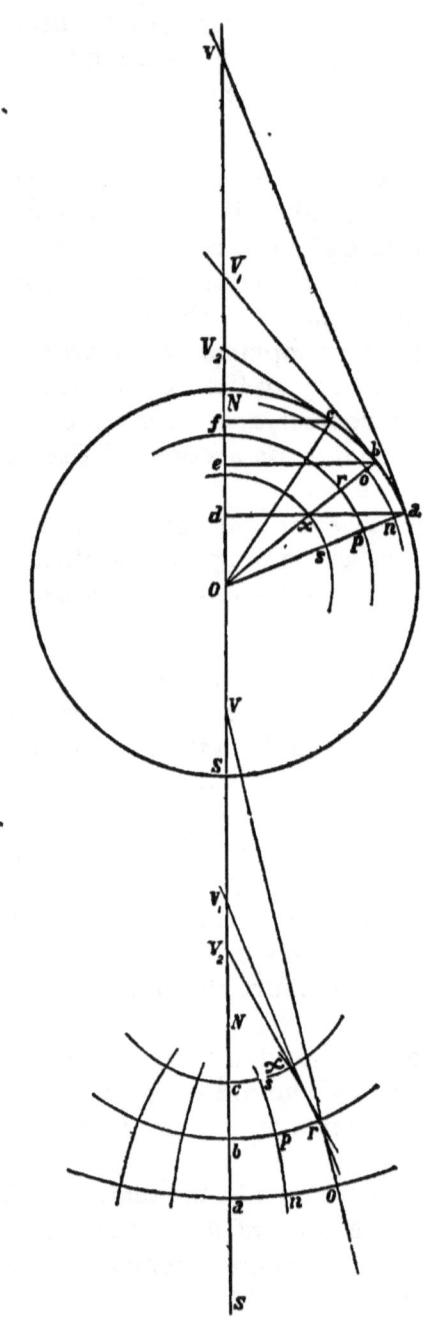

Then in Fig. 2, on the straight line $NS$, make $ab$, $bc$, equal to the arcs $ab$, $bc$, Fig. 1. On the latter figure draw arcs with centre $O$ and radii $ad$, $bc$, $cf$, giving the distances $no$, $pr$, $sx$ on the respective parallels, corresponding to $ab$ on a great circle, as the equator; and lay them off, as indicated by the lettering on Fig. 2, each way from $a$, $b$, $c$.

The *parallels* will now be arcs through $a$, $b$, $c$, with radii $Va$, $V_1b$, $V_2c$, equal to those marked by the same letters on Fig. 1.

The *meridians* will be curved through $o$, $r$, $x$, and $n$, $p$, $s$, etc. $Vo$ will be the developed meridian at $o$ for the cone $V$, and would be that meridian if the map were made on that cone only (233). $V_1r$ is likewise the developed meridian at $r$ for cone $V_1$, and $V_2x$ that at $x$ for cone $V_2$. The actual polyconic curved meridian, $orx$, is evidently tangent to these mono-conic meridians.

For the accurate construction of maps by the polyconic and Mercator's projection, extensive tables are used by the United States Government Bureau of Navigation.

235. A slight modification of the above, and called the *equidistant polyconic projection*, transfers the accuracy of the scale from the parallels, except the centre one of the map, to the meridians. It consists in laying off $ab$, for example, each way on the meridians, from $p$, $r$, etc., taken on $bpr$ as the central parallel, in order to find points through which to draw the final parallels, those shown in Fig. 2 being in such case temporary. This applies to small areas where the meridians are sensibly straight.

236. *Flamstead's projection.*—This differs from the polyconic, in that, in Fig. 2, for example, the parallels through $a$ and $c$ would have been drawn with the same centre $V_1$, correctly used for the true development, $br$, of the central parallel.

The distances on $NS$ and on the parallels are made of their real size, but the meridians and parallels do not intersect at right angles, as they do in the polyconic projection.

# TRIHEDRALS.

237. *The term trihedrals practically means* here the graphical solution of spherical triangles, as will now be explained.

A spherical triangle, $ABC$, Pl. XXIII., Fig. 25, is bounded by three arcs of great circles of the same sphere, and hence having their centre at $O$, the centre of the sphere.

Such a triangle has six parts, its *three sides*, which are measured by the angles $\alpha$, $\beta$, $\gamma$, subtended by them at the centre of the sphere; and its *three angles*, which, considered as plane angles, are as at $A$, for example, measured by the angles between the tangents at $A$, to the arcs $BA$ and $CA$.

238. But join $O$ with $A, B, C$, and we form a spherical pyramid, $O—ABC$, in which the angle at $O$ is called a trihedral, it being formed by the intersections of three planes which include three diedral angles. Now the tangents at any vertex, as $A$, to the sides, as $BA$ and $CA$, are perpendicular to the radius $OA$, and therefore include between them the measure of the diedral angle between $OCA$ and $OBA$. Also the *plane angles*, $\alpha$, $\beta$, $\gamma$, of the trihedral measure the sides $BC$, $AC$, $AB$, of the triangle.

Thus a spherical triangle is the base of a spherical pyramid whose vertex is a trihedral. The plane angles of this trihedral measure the *sides*, and its diedrals measure the *angles* of the spherical triangle.

If the three plane angles are each of 90°, the trihedral is said to be *tri-rectangular*.

239. Any three parts of a spherical triangle being given, its remaining parts can be found. In the graphical solution, all these are represented by the equivalent parts of the corresponding trihedral.

The solution "*in space*" for the six cases thus arising, is summed up for all of them in the simple statement that the construction of the diedrals, equivalent to the *angles* of the triangle, is an application of (D. G., Prob. XVII.). "To find the angle between given planes," and that of the plane angles equivalent to the *sides* of the triangle, is an application of (D. G., Prob. XV.). "To find the angle between given lines."

. Also, in every case designate the sides of the triangle by $\alpha, \beta, \gamma$, and its respectively opposite angles by $A, B, C$.

PROBLEM I.—*Given the three sides of a spherical triangle, that is, the three plane angles of a trihedral; to find its angles, that is, the diedrals of the trihedral.*

Pl. XXIII., Fig. 26. Let one of the plane angles, $\alpha$, be taken in **V**, and with one of its sides perpendicular to **H**. Thus, $O$ being the vertex, and $OC$ one edge of the trihedral, let $\alpha, \beta, \gamma$ be the given plane angles; $Ob''$ being equal to $Ob$, which gives $a''b''$, base of the face $\gamma$. Describe arcs with $C$ as a centre, and radius $Ca''$, and with $b$ as a centre and radius $a''b''$; and they will intersect at $a$, giving $Cba$ for the intersection of the trihedral with **H**, and hence $aCb =$ the required angle $C$. Then, as is obvious on inspection, by (238) $B$ and $A$ are the two other required angles, $Cp$ being perpendicular to $Ca$, as $Cq$ is to $Cb$, and $B''Cq$ being the plane (237) of the angle $B$; where $CB''$ is perpendicular to $Ob$ as $CA'''$ is to $Oa''$.

EXAMPLES.—1°. Let one edge of the trihedral be perpendicular to **H**.
Ex. 2°. Let the angle $\beta$ be in **V**.

272   TRIHEDRALS.

PROBLEM II.—*Given two sides and the included angle of a spherical triangle, that is, two plane angles and the included diedral, of a trihedral; to find the remaining parts.*

Pl. XXIII., Fig. 27. Let $\alpha$ and $\beta$ and $bCa$, $= C$, be the given plane, and included diedral angles.

Revolve $COa''$ about $CO$ till $Ca''$ takes the position $Ca$, then find $O''$ by intersecting arcs with $a$ and $b$ as centres, and $Oa''$ and $Ob$ respectively as radii. Then $aO''b = \gamma$.

$CAp$, $= A$, and $CBq$, $= B$, are then found as in the last problem.


PROBLEM III.—*Given two sides and an opposite angle of a spherical triangle, that is, two plane angles and the diedral opposite one of them, in a trihedral; to find the remaining parts.*

Pl. XXIII., Fig. 28. Let $\alpha$ and $\beta$ be the given plane angles and $CBq$ the diedral opposite $\beta$, and located as before by revolving its plane $qCB''$, perpendicular to $Ob$, into **H**. Then $bq$ determines the position of the base $ab$ of the face $\gamma$ of the trihedral, which base is limited at $a$ by the arc $a''a$, representing as before the revolution of the face $Oca''$ about $OC$, to its primitive position $OCa$. This gives $bCa = C$, after which $A = CAp$, is found as before. $a'''b = ab$ and $Oa''' = Oa''$, giving $a'''Ob = \gamma$.


240. The three remaining cases of the spherical triangle, or of the equivalent trihedral, may be solved as independent problems by constructions similar to the one just given; or, by the principle of supplementary spherical triangles, they may be resolved into the three just given.

*By the former method* it will be useful to remember that the planes of the diedral angles measured by $A$ and $B$, Fig. 25, are both perpendicular to the face $OAB$; hence, if passed through the same point on $OC$, they will intersect each other in a common perpendicular from that point to $OAB$, and hence to their traces on $OAB$. Now in Figs. 26, 27, 28, $Ap$ and $Bq$ are the revolved positions, in **H**, of just such traces, since $CB''$ and $CA'''$ are both drawn through $C$. Hence the perpendiculars from $C$ to $Ap$ and $Bq$ are the transposed positions of this perpendicular, and are therefore equal, or $Ap$ and $Bq$ are both tangent to an arc having $C$ for its centre and either of these perpendiculars for its radius.

241. *By the second method*, if, for example, the three *diedral* angles $A$, $B$, $C$, are given, their supplements will be the *plane* angles $\alpha$, $\beta$, $\gamma$, of the supplementary trihedral, which returns to Problem I. Then finding the $A$, $B$, $C$ of this supplementary trihedral, their supplements will be the required plane angles $\alpha$, $\beta$, $\gamma$, of the given trihedral.

242. Likewise for the other two cases; having given, for example, $A$, $B$ and the included plane angle $\gamma$, resolve this into Prob. II., by taking the supplements of $A$ and $B$ and $\gamma$, which will be the $\alpha$, $\beta$, and $C$ of the supplementary trihedral, in which find $A$, $B$, and $\gamma$, whose supplements will be the required $\alpha$, $\beta$, and $C$ of the given trihedral.

EXAMPLE.—Solve the three cases above indicated.

# AXONOMETRIC AND OBLIQUE PROJECTIONS.

243. These special kinds of projection afford a class of figures having more or less perfectly the effect of true perspectives, while they are much more easily made and are thus highly useful, especially in their application to small and mostly rectangular objects, or to objects drawn on a small scale, as in text-book illustrations of machines, apparatus, etc.

244. The principal lines of most buildings, machines, and structures are generally in three directions: one vertical, and the two others horizontal and at right angles to each other.

Such objects are fully represented by their perpendicular projections (6) on three planes, each parallel to two of the directions described, that is, upon a horizontal plane, and on two vertical planes at right angles to each other, together with sections formed by cutting the bodies by other planes, usually parallel to one or more of the principal planes; the portions in front of such cutting planes being removed.

245. The figures thus produced are, however, entirely destitute of the pictorial or perspective effects which would

AXONOMETRIC AND OBLIQUE PROJECTIONS. 275

render them much more intelligible to those who are unfamiliar with descriptive geometry, that is, with the method of projections; while it is comparatively difficult and tedious to construct true perspectives. Hence the great utility, to those who are unfamiliar either with ordinary projections or perspective, of the special and rapid methods, now to be described, which give pictorial effects, more intelligible than plans, elevations, and sections, and are more easily made than true perspectives.

246. *Axonometric projection* is a *perpendicular projection*, made upon a plane which is oblique to all three of the lines described in (244)—that is, to all the edges of a solid right angle; otherwise called a *tri-rectangular trihedral* (238). Also the object is so placed that its vertical lines will appear as such in the figure.

The pictorial effect is thus due to the position of the object, in being placed so that all its dimensions are oblique to the plane of projection, and therefore represented as in perspective, by lines not at right angles to each other.

247. *Oblique projections* are, on the contrary, those in which it is the *projecting lines* indicating the direction of vision (5—6) that are oblique to the plane of projection.

Thus, Pl. XXIV., Fig. 10, if $AB—A'B'$ be a perpendicular to the vertical plane, as given by its common or perpendicular projections, and if then $A'a'$, equal to $AB$, the true length of the given line, be called the projection of this given line, it shows that the lines which project the points of $AB—A'B'$ upon **V** have changed, from being perpendicular to **V** to being oblique to it, at an angle of 45°, as at $Aa—A'a'$, where the true size of the angle $AaB$, found by (Prob. XII.), will be seen to be 45°.

The line $A'a'$, equal to $AB$, may extend in any direction from $A'$. Also $A'a'$, instead of being equal to $AB$, may be any simple part of it, as one half, or one third, etc.

248. Thus in two ways oblique projections yield pictorial effects; first, by the many *directions* in which the pro-

jection of a given perpendicular to the plane of projection may lie ; second, by the various ratios between the length $AB$ of the given line and its projection $A'a'$.

The following problems will sufficiently illustrate both of the foregoing kinds of projections, or perspectives, as they are sometimes called from their pictorial effect.

PROBLEM I.—*Having given three axes, the axonometric projections of the three edges of a solid right angle, to find the scales of those edges—that is, the ratio of distances on them to their true size.*

Pl. XXIV., Fig. 1. Let $O$ be the projection upon the paper, of the vertex of a tri-rectangular trihedral or solid right angle, and $Oa$, $Ob$, $Oc$, drawn at pleasure, and of indefinite length, the projections of its edges. As $O$ may be at any height from the paper, we may assume $a$ as the trace of $Oa$ upon the paper. Then as each edge is perpendicular to the plane of the other two (246), $ab$, perpendicular to $Oc$, and $ac$, perpendicular to $Ob$, are (Theor. II.) the traces of the planes $aOb$ and $aOc$ respectively upon the paper. The trace $bc$, of $bOc$, will then be perpendicular to $aO$.

The lines $Oa$ and $Ok$ in space form a right angle at $O$, which is inscribed in a semicircle having $ak$ for its diameter, hence make $XY$, parallel to $ak$, the ground line of a plane perpendicular to the paper, project $ak$ upon it at $a'b'$, on which describe a semicircle, and project $O$ upon it at $O'$, when $a'O'b'$ will show the angle mentioned, and $O'a'$, the true length of $Oa$ and its inclination $O'a'o$ to the paper.

Revolve $Ob$ and $Oc$ about a perpendicular to the paper at $O$, till they are, as at $Ob_1$ and $Oc_1$, parallel to $XY$, when their projections, $O'b_1'$ and $O'c_1'$, will show likewise their true length and obliquity to the paper.

Next, take any distance $a'm_1 = b_1'n_1 = c_1'r_1$; and their projections, $a'm$, $b_1'n$, $c_1'r$, on $XY$, will be the lengths of $a'm_1$ in the directions of the three axes $Oa$, $Ob$, $Oc$, respectively.

Or, to take a particular case, let the scale of the plans

AXONOMETRIC AND OBLIQUE PROJECTIONS. 277

and elevations be one of two inches to a foot, then if $a'm_1$, etc., = 2 inches, $a'm$, $b_1'n$, $c_1'r$ will be scales of 2 inches to 1 foot on $Oa$, $Ob$, $Oc$, respectively; as indicated on the same lines shown separately below.

Fig. 2 repeats the same construction, varied by a difference in the relative directions of the axes, and in the scales upon them.

PROBLEM II.—*To construct the axonometric projection of a monument embracing prismatic and pyramidal portions.*

Let the scales be those of Fig. 1.

Pl. XXIV., Fig. 3. Suppose the base to be a square each side equal to $a'm_1$, Fig. 1. Then make $An$ and $Ar$ equal to the scales $Ob,n$ and $Oc,r$. $AB$, the height of the base, is the projection on $XY$ of its real height $a'B$, shown at $OB$ and $OB'$ on scales $a'm_1$ and $Oa$ respectively.

To find $y$, make $b_1'e_1 = c_1'f_1$, Fig. 1, the true perpendicular distance between the vertical planes $ABC$, $ABD$, and their parallels $GFy$ and $HFy$, and lay off $Be$ and $Bf$ respectively equal to $b_1'e$ and $c_1'f$, Fig. 1, the proper scale values of these distances. Also make $Ey$, the altitude of the sloping portion of the base, equal to $a'y$, the value on the scale $Oa$, Fig. 1, of the real altitude $a'y_1$. Then draw $yg$, limited by making $Ce = Be$ and drawing parallels $cI$, to $eE$, and $Ig$, to $Ey$. In like manner, $h$ is found.

The whole drawing is completed by operations similar to the foregoing, which have illustrated all the scales. All the altitudes $Fy$, $Vs$, etc., are from the scale $Oa$, $= a'm_1$, Fig. 1; all the right-hand widths, as $st$, half width of base of the pyramid, from the scale $ob = b_1'n$, Fig. 1; all the left-hand widths, $yh$, $sb$, etc., from the scale $Oc = c_1'r$, Fig. 1, and all the *real* dimensions from scale $a'm_1$, Fig. 1. Thus, $Vs = a'v$ projection of the real altitude $a'v_1$, Fig. 1, upon $XY$.

249. When two of the three axes are equally inclined to the plane of projection, the projection is called mono-

dimetric. The third axis then makes equal angles, in projection, with the other two. There are two general cases; in one, both of the equally inclined axes are horizontal on the object, (244) in the other, one of them is vertical, and the other horizontal.

Very particular cases are those in which one of the three axes is parallel to the paper. In these cases the plane of the other two is perpendicular to the paper, and they are thence projected in one line. In this case, none of the *horizontal* surfaces of the monument, for example, would be seen, and the pictorial effect would be less striking.

250. *Isometric projection.*—When all three of the axes are equally inclined to the plane of projection, the figure becomes the familiar isometric projection. Pl. XXIV., Fig. 4. In this case, the axes are also equally inclined to each other, therefore at angles of 120°.

251. *Isometric scales and protractors.*—Let $aC$ and $OC$, Pl. XXIV., Fig. 4, be drawn to make angles of 45° with $Oa$. Then the circle of radius $Ch$ will represent in its true *form*, and on the scale of the isometric figure, the circle inscribed in the upper base $aObs$ of the prism. Then divide the quadrant $e,g_1$ into any desired number of equal parts, and produce the radii through the points of division, to meet $Oa$. Then $o$ being the isometric projection of $C$, these radii become transformed into their isometric radii; $Ca$, $Cn$, etc., into $ao$, $no$, etc. Then if each of the latter be equally divided into the same number of parts, it will form a scale for all lines parallel to it.

Again: draw $e,e$ perpendicular to $Oa$, and $eH$ parallel to $Ob$. Then $F$ and $H$ on $oa$ and $os$ will be extremities of the semi-axes of the ellipse which is the isometric projection of the inscribed circle just mentioned. The intersections of this ellipse with $ao$, $no$, $ho$, etc., will then form an angular protractor; in the figure one measuring isometric angles of 15° each.

252. These intersections, as $p$, $q$, $r$, can also be found as follows. The semicircle on $EF$ as a diameter shows in its true *size*, and *parallel to the paper*, as well as in its circular form, the semicircle whose projection is the semi-ellipse $EHF$. Then divide each quadrant of it, as $Fm$, into six equal parts; as before, at $e_1g_1$, so now at $p_1$, $q_1$, $r_1$, etc., and draw perpendiculars $p_1p$, $q_1q$, $r_1r$, etc., to $EF$. These perpendiculars represent the arcs in which $p_1$, $q_1$, $r_1$, etc., revolve back to their isometric positions, and they will meet the isometric ellipse $EHF$ in the same points in which they meet the isometric radii $hoq$, $op$, $or$, etc.

PROBLEM III.—*To construct the axonometric projection of a vertical cylinder capped by a hemisphere of less diameter.*

Pl. XXIV., Fig. 5. We will use the scales of Fig. 2, where $as_1$, the scale of plans and elevations, may now represent a scale of *three feet to an inch*, and scales $Oa$, $Ob$, $Oc$, the same properly reduced, as in Prob. I., to the axes similarly lettered.

Then the radius $OR$, $= or$, parallel to $Ob$, Fig. 2, is 4′ : 4″ taken from the scale $Ob$; $OS$, parallel to $Oc$, Fig. 2, is the same, from scale $Oc$, and the altitude $Oo$ of the cylinder is 5′ : 2″ from scale $Oa$.

$OG$, the real radius parallel to $bc$, Fig. 2, of the sphere, and equal to $OF$, is 3′ : 6″, and is laid off on $OR$ and $OS$, as taken from scales $Ob$ and $O$ respectively.

$OQ$ corresponds to $Or$, projected in $O'c'$, Fig. 2; hence laying off $c'D$, Fig. 2, equal to $OD$, its projection $c'Q$ gives the semi-conjugate axis $OQ$, Fig. 5; and $OP$ may be similarly found.

We can now draw the base, $GPE$, of the hemisphere, and its circular contour, $EFG$, of radius $OG$.

*Any vertical semicircles of the hemisphere* may be found as follows:

*First.* The one on the diameter $HK$ parallel to $Ob$, Fig. 2, and in a plane parallel to $aOb$, Fig. 2. Its vertical radius

$ON$ is the distance corresponding to the true size $OF$, taken from the scale $Oa$. Its semi-transverse axis is $OF$. Its semi-conjugate $OT$, perpendicular to $OF$, and hence parallel to $mOc$, Fig. 2, may be found from the projection of $Om$ (after revolution till parallel to the side plane on $XY$) as $OQ$ was found from $Or$. Or, $OT$ may be thus found: With centre $K$ and radius $OF$, describe an arc cutting $TOS$ at $s$; then $sKt$ gives $Kt =$ the desired semi-conjugate $OT$, by the property of the ellipse that if a straight line move so that two points of it at a fixed distance apart move on two fixed intersecting straight lines, any other point of the moving line will describe an ellipse.

The required semi-ellipse $HNK$ can now be drawn.

*Second.* Any vertical semicircle, as that on the diameter $LM$. The bases of the cylinder being parallel to the plane $bOc$, Fig. 2, draw $Oy$, Fig. 2, parallel to $LM$; then $ay$ is in the plane of the paper, and hence of its real size; hence, draw $Oy$ parallel to $ay$, Fig. 2, and $y$, its intersection with $EFG$, will be an extremity of the semi-transverse axis of the required elliptical projection of the proposed semicircle on $LM$. Then $Oz$, its semi-conjugate axis, can be found as shown, as $OT$ was; or it can be found, as $OQ$ was, by drawing a perpendicular from $O$ to $ay$, Fig. 2, and proceeding as described for $OT$, as found from $Om$, Fig. 2. The semi-ellipse $LzNyM$ can therefore be readily sketched.

EXAMPLES.—1°. Let $Oo$ lie in the direction $Ob$ or $Oc$.

Ex. 2°. Substitute for the given body a niche.

Ex. 3°. Replace the cylinder, Fig. 5, by a cone with its axis successively in the directions $Oa$, $Ob$, $Oc$.

PROBLEM IV.—*To construct comparative views of one given object in elevation, axonometric, and oblique projections.*

Pl. XXIV., Figs. 6—9. Let the object be a conical frustum, $EFef$, mounted on a cylindrical axis.

Fig. 6 represents this object so as to be considered either as a plan or an elevation, as both views would be alike if the axis $AB$ were parallel to the ground line. This figure shows the lengths and diameters in their true size, but except by its title or an end elevation, could not, even with plan and elevation, be certainly distinguished from the frustum of a square pyramid mounted on a square prism having the same linear axis, $AB$.

Fig. 7 is an axonometric projection, made from the real dimensions afforded by Fig. 6, by using the scales of Fig. 2. Here $Cf$, parallel to $ac$, Fig. 2, is the true size of the radius $Cf$, Fig. 6, shown on the scales $Oa$ and $Oc$, Fig. 2, at $Ch$ and $Cg$. Similar explanations apply to the remaining parallel circles. The parts of $AB$ are those of $AB$, Fig. 6, reduced to the scale $Ob$, Fig. 2.

Fig. 8 is an oblique projection of the object when its axis is perpendicular to the paper at $A$, and the projecting lines make an angle of 45° with the paper, in a plane whose trace on the paper is $AB$, thus projecting the point $B$ at $B$ so that $AB = AB$, Fig. 6.

The circles of the object thus being parallel to the paper, are projected as circles in their real size, and the figure is as easily made as Fig. 6. It has a somewhat lengthened appearance, because $cf$ is greater than its real size $cf$, Fig. 6; but this could be remedied by making the parts of $AB$ shorter, in a uniform ratio, than the like parts of $AB$, Fig. 6. This modification supposes the projecting lines to make some angle greater than 45° with the paper. The circles would still remain circles of their present size.

Fig. 9 represents the same object as before, but with the axis $AB$ parallel, and the planes of the circles perpendicular to the paper. It illustrates the considerably increased complexity of construction with more distorted result than in Fig. 7 or 8. The diameters, as $2Cf$, are here parallel, and those represented by $pp_1$, are perpendicular to the paper, but as $Cp = \frac{1}{2}Cf$, the projecting lines make a larger angle than 45° with the paper.

Describing a circle on $2Cf$ as a diameter, draw the ordinate $CP$, and representative $Pp$, of a projecting line. Thus the radius perpendicular to the paper at $C$, and revolved at

$CP$, is transformed into the oblique projection $Cp$, and the line $Pr$ into the line $rp$, since $r$ in the paper is its own projection, and $p$ is that of $P$.

Reasoning thus, draw the diameter $tq$ perpendicular to $Pr$; the tangent at $t$ to the circle $CP$, and parallel to $Pr$, will be transformed into a parallel to $rp$ at $v$, to which the elliptical projection of circle $CP$ will be tangent, as at $a$.

Finding $q$, $m$, and other points, and the opposite tangent at $x$, parallel to that through $v$, by similar constructions, or making $Cp_1$, etc., $= Cp$, etc., the elliptical projection of circle $CP$ can be sketched. The remaining circles can be similarly found. The parallel tangents to the small circles are found as that at $v$ is, and are at more than their real distance apart. The contour of the conical part consists of tangents to its bases.

Comparing Figs. 8 and 9, the advantage of making circular outlines parallel to the plane of the paper is obvious.*

EXAMPLES.—1°. Construct Fig. 7 with $AB$ either in the direction of axis $Oa$ or $Ob$, Fig. 2.

Ex. 2°. In Fig. 8 let the segments of $AB$ be made two thirds, or one half their real size.

Ex. 3°. Make an isometric projection of Fig. 6.

* See my Elementary Projections and Shadows, or my Elementary Projection Drawing.

www.ingramcontent.com/pod-product-compliance
Lightning Source LLC
Chambersburg PA
CBHW030820230426
43667CB00008B/1307